T0056026

The Making of Modern Medicine

THE MAKING OF MODERN MEDICINE

Turning Points in the Treatment of Disease

+++ **MICHAEL BLISS** +++

THE UNIVERSITY OF CHICAGO PRESS *Chicago and London*

Michael Bliss is University Professor Emeritus at the
University of Toronto. His books include *Plague:
A Story of Smallpox in Montreal*; *William Osler:
A Life in Medicine*; *Harvey Cushing: A Life in Surgery*;
The Discovery of Insulin; and *Banting: A Biography*.

The University of Chicago Press, Chicago 60637
The University of Chicago Press, Ltd., London
© 2011 by Michael Bliss
All rights reserved. Published 2011
Printed in the United States of America

20 19 18 17 16 15 14 13 12 11   1 2 3 4 5

ISBN-13: 978-0-226-05901-3 (cloth)
ISBN-10: 0-226-05901-4 (cloth)

Library of Congress Cataloging-in-Publication Data
Bliss, Michael, 1941–
The making of modern medicine: turning points in the
treatment of disease / Michael Bliss.
p. cm.
Includes bibliographical references and index.
ISBN-13: 978-0-226-05901-3 (cloth: alk. paper)
ISBN-10: 0-226-05901-4 (cloth: alk. paper)
1. Medicine—Western countries—History—19th century.
2. Medicine—Western countries—History—20th century.
3. Medical innovations—Western countries—History—19th
century. 4. Medical innovations—Western countries—
History—20th century. 5. Smallpox vaccine—History.
6. Insulin—History. 7. Medical education—Western
countries—History—19th century. 8. Medical education—
Western countries—History—20th century. I. Title.
RI49.B55 2011
610—dc26   2010014691

⊗ The paper used in this publication meets the minimum
requirements of the American National Standard for
Information Sciences—Permanence of Paper for Printed
Library Materials, ANSI Z39.48-1992

Contents

Acknowledgments · vii

Introduction · 1

I ✳ Fatalism: Montreal, 1885 · 5

II ✳ The Secular Saints of Johns Hopkins · 31

III ✳ Mastery: Toronto, 1922 · 63

Epilogue: The Collapse of Life Expectancy · 87

Notes · 95   *Index* · 101

Acknowledgments

Thanks to the Department of History at the University of Western Ontario for inviting me to be the speaker for the 2008 Goodman Lectures, and to everyone at Western Ontario who made giving the lectures a pleasant and educational experience, especially two of my former students, professors Ben Forster and Shelley McKellar. Thanks to Marian Hebb in Toronto and Karen M. Darling at the University of Chicago Press. As always my greatest debt is to the person who has sustained me throughout all the vagaries of everything, my wife, Elizabeth. And now I am writing for four special readers, our grandchildren Kate and Michael (the chicken pox kids), Jasmin and Joey.

INTRODUCTION

This book is about the coming of age of modern medicine between 1885 and 1922 in what were then some of the most advanced Western countries: the United States, Canada, and, to a
degree, Great Britain. It traverses a period that starts with physicians' helplessness in the face of an epidemic of a terrible infec-
tious disease, and ends with medical researchers' ability to spec-
tacularly alter the human condition by minimizing the killing
power of a previously deadly disease. It is also an interpretation
of the research in the history of medicine that I have published
during a varied career as an academic historian.

I was trained to be a professional historian, not to be a doctor
or a scientist, although science was one of the first of my academic
pursuits. My second pursuit was philosophy, with a special interest in the history of religion. My third interest was the history of
my native country, Canada, in virtually all of its dimensions —

economic, social, and political—and particularly during the years from roughly the 1870s to the 1930s, which seem critically formative of the modern society in which I and my students make our lives. In about the middle of my career, all of these interests began to coalesce in what for me was a kind of return to the family business, medicine. My father had been a small-town general practitioner, and a deceased older brother had been a medical researcher. As a child I had wanted to be a doctor like my father. Although in a teenage epiphany I had turned my back on medicine as an academic career, in the late 1970s I wandered back to medicine.

I have written five previous books in interrelated areas of medical history. Some of the interrelatedness includes the use of Canada as a backdrop; it also involves my interest in some of the philosophic and religious dimensions of my subject matter and the late nineteenth- and early twentieth-century time frame in which I have felt most at home. On the other hand, read individually, my medical history books appear to be distinct explorations of quite different topics—the discovery of insulin, a ghastly smallpox epidemic in one city, and biographies of a high-achieving researcher, an outstanding physician, and a pioneering surgeon.

Gradually I realized that if an explorer ventures into terra incognita from several directions and does enough wandering around, he eventually accumulates a reasonably coherent knowledge of his terrain. My territory and my explorations were relatively tiny contributions to the mapping of the global history of healing, a vast challenge for generations of medical historians, but my work has dealt with some fairly big and important issues, people, and discoveries. In 2000–2001 I made a first attempt at an overview of the themes of my medical histories.[1] Some years and another big book later, plus retirement from my professorial teaching responsibilities, the opportunity to deliver the 2008 Joanne Goodman Lectures at the University of Western Ontario prompted me to return to the effort of synthesizing and summing up. This little book, a revised and expanded version of my Goodman Lectures, is the product. It's a synthesis of what I have tried to say as a historian of medicine.

My method is to use three case studies—three case histories—to illuminate what was probably the key turning point in our attitudes about modern health care, a turning point that took place near the end of the nineteenth century and the beginning of the twentieth century. This was the time in history when our trust in the capacity of physicians to actually treat disease became established.

Second, in a very broad way I am meditating here on how certain concepts of "faith" have changed with the development of modern health care. These studies show how the evolution of our healing capacities can be situated in the context of a shift from humanity's age-old hope for supernatural healing and salvation—faith in the old gods—to a new faith in the capacities of secular caregivers to offer us limited but extremely desirable salvation from the ravages that disease and time inflict on our bodies. The growth of this faith seems to me to be a central theme of the rise of modern medicine. It is one that we ought to better understand in present times when we are in danger of becoming bogged down in despair about the seeming paradox of ever-mounting health care costs in an age of unprecedentedly good health.

Third, these studies reflect my interest in the relative prominence that many aspects of North American medicine came to enjoy, compared with other countries. The twentieth century was the American century in many ways, not least in science and medicine, as the United States became the leading global center for medical research, medical innovation, medical technology, and pharmaceutical development, as well as advanced clinical and hospital practice. It did not become the global leader in health insurance, but that is another story. Canada, a much smaller North American country in terms of population, participated significantly in the North American medical coming of age. Indeed, one argument made here is that Canada played a major role as a catalyst in American development by serving as a conduit helping to bring important European attitudes toward medical education and research to the United States.

Throughout the twentieth century, Canadians liked to believe that they punched above their collective weight in most areas of health care (including health insurance), and perhaps they did. At the very least, the Canadian contribution in giving the world William Osler and insulin therapy for diabetes stands in good comparison with practically any other milestones made anywhere in the history of healing. Incidentally, focusing on the interplay of various aspects of the triangular relationship between Canada, the United States, and Great Britain—what has sometimes been called the North Atlantic Triangle—has been one of the stated themes of the Joanne Goodman Lecture series. Funded by the Goodman family to commemorate the brief life of a joyous and promising student of history, for more than thirty years the Goodman Lectures have led to significant publications of interpretations of a variety of topics in North Atlantic diplomatic, political, and social history.

As a guide to the coming of modern medicine and to some of the intellectual and geographical foundations of that development, I offer these three studies of turning points in humanity's assault on disease: the first of a devastating smallpox epidemic in Montreal; the second of the rise of a great temple of healing in the United States and the doctors who personified its aspirations; and the third of the discovery of insulin, one of the first of the miracle therapies that seemed to legitimize the transference of faith from priests to physicians. In an epilogue, I reflect on some of the implications and ironies of these events for health care and the attitudes we hold toward it at the beginning of the twenty-first century.

* I *

FIGURE 1

William Osler. High priest of the coming of modern medicine.

Osler Library, McGill University

FATALISM: MONTREAL, 1885

We spend most of this chapter in late nineteenth-century Montreal, then the dominant city in the Dominion of Canada. It was a commercial, financial, and industrial center of about 165,000 people, surrounded by suburban villages, situated on the St. Lawrence River in the province of Quebec. Sixty percent of Montrealers were of French descent, 40 percent of British. The scene opens in 1874 as a young Montreal doctor, William Osler—the most recently appointed and probably best-trained professor at Canada's best medical school, McGill University's Faculty of Medicine—finds himself confronted with one of his first patients, an English visitor to the city. Osler realizes that the sick man has a case of hemorrhagic or black smallpox, what the French-Canadians call *la picotte noir*.

It is a terrible disease, which Dr. Osler is helpless to cure. He can help his patient a little by prescribing morphine to relieve his

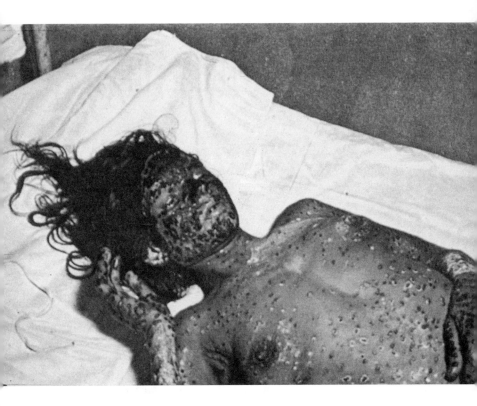

FIGURE 2

Smallpox victim. God's punishment?

World Health Organization

suffering. Osler can also be very confident of both the accuracy of his diagnosis and, unfortunately, the accuracy of a prognosis that holds out little hope. As the victim worsens, Osler stays by his bedside, for hours on end, and at the patient's request he reads to him from the Bible. The man occasionally mutters prayers, and his doctor helps in one further way: "As the son of a clergyman . . . I performed the last office of Christian friendship I could, & read the Commendatory Prayer at his departure"—an act of both priestly caring and medical resignation.[1]

∗

Smallpox was one of the most dramatic and loathsome contagions that physicians like Osler confronted in the nineteenth century. It was far from the only disease that highlighted their helplessness. They had no weapons against bubonic plague, cholera, diphtheria, typhus, typhoid fever, scarlet fever—indeed any kind of infection. They had no weapons to confront any kind of chronic disease, ranging from congenital conditions, such as the deficiency that caused cretinism in children, to systemic breakdowns, such as the onset of diabetes mellitus, the growth of tumors, the decay of the cardiovascular system, and many other afflictions. They were powerless to affect humanity's inevitable progress toward death, as powerless as King Canute was to control the tide, as powerless as we are today to improve our climate.

In such situations, with little to do other than try to make patients as comfortable as possible, it is not remarkable that both physician and patient would fall back on age-old religious rituals, such as the reading of Scripture and the saying of prayers. It is not remarkable that victims stalked by deadly diseases comforted themselves as best they could with fatalistic acceptance, variations on the theme that whatever could not be changed by human activity must be a product of divine will. For Christians, the overwhelming majority of the population of English-speaking countries throughout the nineteenth and twentieth centuries, the good news was that their faith included the prospect that physical suffering and death were followed by a glorious and eternal afterlife. You would meet your loved ones again on the other side of the grave. And so, in times of epidemic and grief, one could at least pray for the souls of the dead, that the God who had taken them would favor them with his grace.

Of course, one also prayed that God might stay his judgment and not inflict disease on his people. If disease had divine origin, and if it struck unevenly, surely there were elements of divine choice at work in the incidence of sickness. Perhaps God was striking down sinners by sending plagues—plagues on Egypt, plagues on his enemies, plagues on the sinful, plagues on the corrupt. Perhaps the coming of disease could be prevented by prayer, by faith, by devotion, by the intercession of the saints, by godliness, by avoiding sin, by taking what might be called religious precautions—always remembering that the ways of the Lord were ultimately unfathomable, so there could be no guarantee of exactly how prayers and devotions would be answered, if at all.

By the last quarter of the nineteenth century, the continuing development of humanist, secular, empirical, and scientific explanations for the onset of disease conditions mounted a powerful challenge to the predominantly or fundamentally religious worldview. That challenge was no more evident than in the view that sanitation, both public and private, could obviate the conditions of filth, odor, miasma, and pollution that seemed to give rise to so many diseases, the exact causality of which was usually unknown. For many people in the Western world, cleanliness had begun to rival godliness—even displace godliness—as a precondition of health. Beginning in England in the 1830s, public health through improved sanitation had become a great crusade in the age of progress. Sanitation did prove vitally important in helping prevent the spread of filth- and waterborne diseases, such as the greatly dreaded cholera.

It had also happened that one of the greatest breakthroughs in the history of health had occurred at the end of the eighteenth century in connection with the equally dreaded and possibly even more loathed epidemic disease, smallpox. One of our most familiar medical stories, broadly true in its essential details, is of how an English country doctor, Edward Jenner, discovered in the 1790s that if he inoculated people with the germs of cowpox— that is, if he gave them *variolae vaccinae*, smallpox of the cow— they would develop an immunity to human smallpox. Exactly how this happened would be a mystery for many years, but that it occurred offered marvelous possibilities. Prevention by vaccination almost immediately replaced the only other preventive procedure for smallpox, which had been inoculation with the human virus itself. At best this inoculation induced a mild case of smallpox and subsequent immunity. Too often it would ignite what could become a smallpox epidemic.

Vaccination was quickly heralded and adopted around the world. In 1798, the same year Jenner published his landmark pamphlet on vaccination, a teenage boy in far-off Newfoundland was vaccinated successfully by one of Jenner's former students. By the 1820s, millions of people had been vaccinated in Europe and America, the pope had endorsed vaccination as a gift from God, and humanity stood on the threshold of salvation from one of its most-feared scourges. As Osler knew in 1874, physicians were still as helpless to treat actual cases of smallpox as they were of most other diseases. But here was a disease that could fairly easily and usually reliably be prevented. In the first two-thirds of the nineteenth century, the incidence of smallpox outbreaks in Western countries decreased rapidly, becoming unusual occurrences controllable by quarantine and vaccination. Foresighted public health workers speculated, as had Jenner himself, that

FIGURE 3

Vaccination, 1873. Historic breakthrough; hope for humanity.

World Health Organization

smallpox was a disease that in theory could be completely con-
quered, completely eradicated.[2]

*

On February 28, 1885, George Longley, a conductor on a Grand
Trunk Railway Pullman car that had arrived in Montreal from
Chicago, felt he was coming down with something. A local doctor diagnosed a mild case of smallpox and, after various misad-
ventures (relating to the fact that the conductor was a Protestant
but the Montreal General Hospital would not treat cases of smallpox), had Longley admitted to the Roman Catholic Hôtel-Dieu
Hospital. Perhaps because of some early confusion about whether
or not Longley had smallpox or chicken pox, Hôtel-Dieu's procedures for isolating the highly contagious patient broke down.
Although Longley fully recovered from his case, the disease
spread to hospital workers, two of whom died.

A completely separate hospital building, the city's old smallpox hospital (it had been busy in the 1870s when smallpox was
endemic in Montreal but had been closed for the past several
years) was reopened so the contagious patients could be removed
from Hôtel-Dieu. But this action came too late. As smallpox continued to spread in Hôtel-Dieu, it was decided in mid-April that
the whole building had to be disinfected. All patients who were
able to be relocated and did not seem to have smallpox were sent
home. Almost immediately cases of smallpox outside of the hospital began to be reported. The civic Board of Health realized there
was a serious epidemic of smallpox in the streets of Montreal.

The health authorities thought it could easily be contained
and the disease stamped out. All patients were to be taken to the
smallpox hospital, and their residences were then to be sealed
and disinfected with sulfur fumes. In rare cases where patients
could not be removed, their residences were to be isolated and

the inhabitants quarantined. In the meantime, vaccination was to be made freely available. Montrealers who did not want the services of the public vaccinators could be vaccinated by their private physicians. By the end of April, it was thought that the mini-epidemic, which had taken only about six lives, was over.

In fact, everything was going wrong. Complaints began to pour in to the Board of Health about harmful effects of vaccination, notably the development of cases of erysipelas (skin inflammation and fever, sometimes acute) among the recently vaccinated. Parents claimed that vaccination was making their children desperately sick, perhaps even killing them. Realizing that its vaccine supply was contaminated, the Board of Health decided to suspend public vaccination, relying instead on private physicians for vaccination and on procedures for removal, isolation, and quarantine. The health officers almost immediately found, however, that in the poor neighborhoods of Montreal, where the flames of smallpox were concentrated, people did not take any preventive or control methods seriously.

They would not cooperate with the health authorities. They would not send their sick children to the smallpox hospital. They would not even isolate their children. When the sanitary police placarded their homes, they tore down the placards. Quarantine was ignored. Children and adults with freshly pockmarked faces were observed in the many crowds that gathered in Montreal that summer—crowds that massed to observe religious processions and a bishop's funeral, to welcome troops home from helping suppress Louis Riel's rebellion in Canada's Northwest Territories, and to attend Buffalo Bill's Wild West Show. Many Montrealers were vaccinated or re-vaccinated by their doctors, but many others ignored the preventive. At least two apparently qualified local doctors spoke out strongly against any kind of vaccination. They claimed that vaccinations were a foul injection of animal disease into human bodies, that they caused cases of smallpox as well as

FIGURE 5A,B

Anti-vaccination broadsheets, Montreal 1885.

Appealing to fear, ignorance, and fatalism.

Fisher Library, University of Toronto

MAD!!

Our City Authorities and Press are **MAD.** Their insane cry of **ALARM!!**
"VACCINATE!"—"VACCINATE!" has driven thousands of our usual
summer visitors away from the city, and injured our trade and commerce to the
extent of millions of dollars by their senseless ravings. Small-pox is not epidemic
in Montreal at present; it is sporadic and in general endemic, but considering the
density and character of our population of 200,000 people of mixed nationalities,
there have been very *few cases*, not exceeding 500 at any time in the city, and these
generally in the most ill-conditioned localities. The last Report of our Board of
Health says :—" Up to date (Aug. 17) there have been **133 PATIENTS** admitted
into the Civic Hospital suffering from small-pox, of these **SEVENTY-THREE
WERE VACCINATED, FIFTY-SIX** had **ONE** mark on the arm,
THIRTEEN TWO marks; and **FOUR THREE** marks. In all **44 DIED**;
of these, **EIGHTEEN WERE VACCINATED.''** There is no reason for this
SENSELESS PANIC. Montreal should be one of the healthiest cities on this
continent, and it would be if our authorities would enforce thorough **SANITATION**
instead of blindly trusting to the **miraculous** potency of a **USELESS, DAN-
GEROUS,** and filthy **RITE** like Vaccination.

Dr. Garth Wilkinson, the eminent English physician and author, stated before
a Committee of the English House of Commons, 1871, that " a small-pox panic in
London is worth *one million pounds sterling* to the profession!" If so, what is our
panic worth ?

OUTRAGE ON PERSONAL LIBERTY!

MONTREAL WORKING-MEN AND WOMEN
FORCED TO BE VACCINATED!!

No work for those who refuse to have the mark of the **BEAST** on their bodies!
TALK NO LONGER OF RUSSIAN TYRANNY! Tyranny is
detestable in any shape, but in none so formidable as when it is assumed and
exercised by a number of petty tyrants. It is in vain for working-men and women
to plead that they do not believe in the efficacy of Vaccination. They are told that
they may believe what they like, but that Vaccinated **THEY MUST BE,** or
leave their employment, which to many of them means **STARVATION!!**

other infections, and that sanitary precautions—cleaning up the city, especially its slums—would be sufficient to defeat smallpox. The outbreak of smallpox was not contained. The red death, as Edgar Allan Poe and others had called it, spread like fire. By mid-August 1885, 120 Montrealers had died from smallpox, and the disease seemed to be raging out of control. The city finally became alarmed, as did the outside world. Several newspapers began giving the epidemic front-page coverage. Many more Montrealers rushed to be vaccinated by their doctors. The Board of Health resumed free vaccination, using a new vaccine. It beefed up its containment efforts, enlarged the smallpox hospital, and considered laying charges against those who violated the sanitary bylaws.

Tourists and business travelers began to shun what was starting to be seen as a plague-ridden community. Merchants across Canada and the United States began boycotting products made in Montreal factories. In the stricken city, a committee of concerned businessmen and public-spirited citizens began to advocate making vaccination compulsory. The death rate rose to seventy-five per week, then to one hundred per week, with no end in sight.

By the end of September, there were stories of smallpox sufferers literally dying in the streets of Montreal, even as more incidents of violent resistance to patient removals and placarding were reported. After a week when a further 238 Montrealers died from smallpox, the city council decided it had to make vaccination compulsory. As soon as that resolution was passed, crowds of angry Montrealers gathered and began stoning and trashing Board of Health offices. The mob ultimately laid siege to city hall, their protests and battles with police mounting to the level of a full-scale riot.

Montreal's mayor, Honoré Beaugrand, supported by the city council, formally requested that the militia be called out to come to the aid of the civil power. More than one thousand soldiers

FIGURE 6

Forced removal of smallpox victims, Montreal 1885.

Harper's Weekly

were mustered to protect civic buildings and civic officials, and especially to stand guard as a new smallpox hospital was constructed in buildings on the provincial exhibition grounds. Within a day or so, the violence was contained, but there was no attempt to implement compulsory vaccination. One soldier died in a gunshot accident while on guard at the smallpox hospital.

Through October 1885, a grim masque of red death raged in Montreal, with 250 to 300 deaths per week within the city and as many more in several poor villages on its outskirts. Black hearses trundled up and down the streets of the city, especially its east end. Authorities vaccinated where they could, removed patients to the smallpox hospital under armed guard, and laid charges against obstructors. The outside world, including American state governments and the government of Ontario, enforced its own quarantine against Montreal. Passengers on trains leaving the city were inspected at the Ontario-Quebec boundary and the Canada-U.S. border. They were required to show vaccination scars and/or vaccination certificates, or accept vaccination on the spot. A market immediately developed in fraudulent vaccination certificates.

In November and December, the death toll gradually declined, though New Year's Eve 1886 featured a pitched battle as Montreal police defended wooden barricades erected to block streets connecting the smallpox-ridden village of Sainte-Cunégonde with the rest of the city. On January 31, 1886, the epidemic was officially declared ended. Many churches held ceremonies thanking God for relieving the city. Embers of smallpox radiated into spring.[3]

✳

In the fifteen months from February 1885 to May 1886, Montreal recorded at least 9,600 cases of smallpox. There were a further 10,300 cases outside the city, mostly in its poor suburbs. In all,

3,164 Montrealers died of smallpox, about 2 percent of the city's population. There were a further 2,600 deaths in Quebec outside the city, mostly in nearby suburbs. These figures almost certainly understate the real tolls because many cases and deaths were unreported or mislabeled.

In the history of North America, there had never been a smallpox epidemic like this in any city. Montreal's smallpox epidemic of 1885 appears to have been the worst scourge of the disease in any industrial city anywhere since perhaps the beginning of the century, certainly in the last third of the century. While sparks of the smallpox virus did set off minor conflagrations in many cities and countries into the 1880s, they were everywhere else met with fierce resistance and preventive measures that stamped them out. In 1884, for example, an outbreak in a rural township in the province of Ontario was defeated by a state-ordered closing of schools and churches, a ban on all public gatherings, strict enforcement of quarantine, and house-to-house vaccination. In 1885 when travelers from Montreal threatened to set off epidemics outside Quebec, every jurisdiction responded forcefully and with success. The worst spin-off epidemic was in Charlottetown, Prince Edward Island, where in six weeks smallpox took fifty-three lives. Here, too, the virus was stopped by compulsory house-to-house vaccination, a closing of schools and churches, and a ban on all public meetings.

How had a disease that doctors knew could be prevented and that was being overcome everywhere else in the Western world been allowed to kill so many people in a large city in progressive Canada? Clearly there had been failure and negligence. Dr. William Osler, who had left Montreal to take up an appointment in the United States in 1884, mentioned the epidemic in his classic textbook, *The Principles and Practice of Medicine*, as "perhaps the most remarkable instance in modern times of the rapid extension of the disease." He attributed the death toll to "a negligence absolutely criminal."[4] What was the source of the negligence?

*

Not surprisingly, almost all of the deaths were of people who had never been vaccinated.* Almost all of these victims, 91.2 percent of Montreal's deaths, were among the French-Canadian population of the city. Almost all the victims, 85.9 percent, were children under the age of ten. The 1885 smallpox epidemic raged predominantly among unvaccinated French-Canadian children. All the subjective evidence indicates that most of these children were from poor families living in congested, often inadequate housing.

Why did poor French-Canadian parents not get their children vaccinated? I cannot stress too highly that the fact of their poverty is not a sufficient explanation, because the very poorest Montrealers did get vaccinated in 1885. This was the largely Irish population of the Griffintown slum, and there are no reports of their objecting to vaccination or of a significant death toll among their children. Poor Irish and Irish-Canadians did get vaccinated and did not die from smallpox. Iroquois Indians on the nearby Caughnawaga reservation were also vaccinated and were virtually smallpox-free; years earlier their forefathers and other aboriginal North Americans had been among the disease's most devastated victims. Poor French-Canadians did not get vaccinated and suffered terribly. Why?

The answer involves a complex mix of perceptions involving fear and fatalism, untempered by knowledge—perceptions that were encouraged in this population by certain of their opinion leaders who ought to have known better.

Today we often fear the side effects of beneficial therapies. We

---

*Vaccination itself only gave limited immunity from smallpox. Thus a few people who had been vaccinated years earlier were again vulnerable. William Osler himself suffered a mild case of smallpox in 1876.

sometimes fear side effects to a fault. It should not surprise us that in the 1880s some people were fearful of the effects of vaccination, worrying that inoculating healthy people with a disease germ might do them serious harm. Perhaps it would cause smallpox; perhaps it would cause some other pattern of diseases. From the beginning of vaccination in the early 1800s, there was a certain amount of fear and resistance. In most populations it was usually slight and did not impede the development of policy for the masses. Many folk had become habituated to inoculation with smallpox before vaccination was introduced as a next and better step. Many others simply did what their leaders told them to.

Unfortunately in Montreal in 1885, there was not yet a tradition of taking vaccination for granted. Unlike the Irish, unlike the French in France, unlike Indians in reserves across Canada, masses of French-Canadians in Quebec had never been subject to the procedure. Apparently they had not even practiced systematic inoculation. As well, there were a number of prominent spokesmen who suggested to French-Canadians that their fears of the effects of vaccination were well-founded. At least two loud and prominent anti-vaccinationist physicians—"doctor" Alexander Milton Ross, who came from Ontario, and a respected French-Canadian doctor, Joseph Emery Coderre—spread the notion that vaccination generated smallpox and perhaps caused other diseases. Ironically, the vaccinophobes' concerns seemed to have solid substance when the city's original public vaccination program had to be suspended because of tainted lymph. It really did cause cases of erysipelas.

These fears were intensified by ethnic tension in Montreal related to the deep cultural and historical fault lines dividing French from English. Some French-Canadians came to believe that vaccination, an "English notion," was being promoted and encouraged by the English as a race weapon against the French. It was possible to construct fantastic conspiracy theories involving the English Dr. Jenner; the English and unilingual Dr. William

Bessey, who was head of the vaccination corps that spread bad vaccine in Montreal; the English newspapers that dominated the city; the English factory bosses who were advocating compulsory vaccination of French-Canadians; the English aldermen who were the most vociferous supporters of vaccination; and so on—even the English-Canadian politicians who were also leading the movement to suppress the half-breed or *métis* uprising in Canada's Northwest Territories and to execute its French-speaking leader, Louis Riel. At the very least, the many expressions of English scorn at French backwardness in resisting vaccination (also scorn at French credulity, ignorance, and superstition) were like rubbing salt in old and deep wounds. The effect was that vaccinophobia was supported and reinforced by anglophobia.

In resisting innovation, the anti-smallpox movement relied on old but still current ideas. The most insidious of these ideas had once been deeply progressive—the notion that public sanitation was the best and most modern way to fight all diseases, including smallpox. If smallpox, apparently like most other epidemic diseases, was generated in filth and spread through miasmas and other odiferous gases, or perhaps in bad water, then there was a perfectly acceptable alternative to vaccination. Public cleanliness—say, the cleanliness and clean air enjoyed by the folk who tended to cluster on the windward western slopes of Mount Royal, who happened to be mostly English—should do the trick.

This was the fallback position held by most of the anti-vaccinationists. They agreed that smallpox was certainly a serious health problem, but argued that it should be tackled by public health campaigns that did not get sidetracked by the harmful panacea of vaccination. You wouldn't get smallpox if you were never exposed to it (as your children probably would if you allowed the sanitary police to take them to the foul pest-house, the smallpox hospital, where so many of them died), and that would not happen if Montreal were properly cleaned up. Unfortunately this belief was simply wrong, resting as it did in ignorance of how the smallpox virus actually spread. The etiology of smallpox was poorly understood by practically everyone at a time when bacteriology was just in its infancy and virology had yet to be conceived.

Anti-vaccinationist sanitarians, like "doctors" Ross and Emery Coderre, also preached a hostility to "doctor-craft" that reflected early nineteenth-century resistance to a profession formerly mired in harsh interventions, such as bleeding, blistering, and purging, as well as the administration of heavy doses of caustic drugs. Better to prevent illness by cleaning up the natural world, better to treat illness by letting natural healing take its course. Midcentury medical sectarians—such as homeopaths (who treated with minuscule doses of drugs) and hydropaths (who claimed water treatments as a universal cure-all)—were unwilling to submit to any kind of priesthood, clerical or secular. They instead advocated something like medical pantheism, a faith in nature and natural cures. (Anti-vaccinationists, who are still with us, would become the original "conscientious objectors" to legislation requiring compulsory immunization.)

When that was not enough, the tendency was to fall back on mild, quasi-medical remedies for *picotte*—herbs, potions, cooling drafts, ointments, aromatics, even Labatt's beer—that would perhaps offer protection and relief. These were frequently advertised and probably widely used during the epidemic. The patent medicine industry in North America was at the peak of its popularity in an age when regular medicine seemed to offer either powerful, discredited drugs or no drugs at all.

Most significantly, there was a much older faith-based view, to the effect that a person would not get smallpox unless God had willed him or her to have it. It was a worldview stressing the complete omnipotence of a supernatural deity: a deity responsible for both the good and the evil in creation, a deity capable of and willing to intervene constantly in the workings of creation, a

FIGURE 7A,B

Patent medicine ads, Montreal, 1885. Profiting from false hope.

Montreal Herald, 1885

wrathful deity who rewarded righteousness and punished sin, a deity who answered prayers. It was a worldview that led many of its adherents, especially some of Montreal's most devout Roman Catholics, to argue that the best and possibly the sufficient way to fight the epidemic was by renewed devotion and prayers for intercession to stay the hand of the sender of the plague. What if Montreal was being afflicted by a plague because of the sinfulness of its people? Had the city fallen prey to secularism? Had it become dominated by hedonism, sensualism, freemasonry, and other vices associated with Protestantism and Anglo-American culture—and specifically with the toboggan slides at Montreal's 1885 Winter Carnival, where young females had shocked black-robed priestly observers as they tumbled down the runs, skirts awry? To avoid being stricken by contagion of the body, it was necessary to confess and repent contagions of the soul. Pray. Pray to the saints, pray especially to Saint Roch, healer of the victims of plagues past, and have faith that prayers would be answered.

These were not beliefs held by the most visible Montreal churchmen, either Protestant or Roman Catholic. By the 1880s many thoughtful Protestants and Catholics had amended their worldviews to incorporate medical interventions, including practices like vaccination, as divinely sanctioned gifts in the struggle against the consequences of human frailty and natural menace. As mentioned, the Roman Catholic Church had long since officially endorsed vaccination. In Montreal in 1885, the church hierarchy, though initially a bit hesitant, eventually repeatedly advised Catholics to resort to vaccination.

For complex reasons, however, the church trod a fine line in its relations with the state and in its relations with health care. Priests were not doctors, so the clergy refused to recommend from the pulpit any specific medical procedure. Even more firmly, clergy absolutely refused to support proposals that religious gatherings, including public and ceremonial professions of faith, should be banned or limited in the hope of restricting the spread of the disease. It seemed unthinkable that denying people witness and access to the rites of their churches could be an effective weapon against disease. No matter how negligent and foolish the families decimated by smallpox might have been, the church could not deny anyone the comfort of its observances and its belief in eternal salvation.

Action by the hierarchy was also complicated by the strength of a particularly devout strain of Catholic religiosity in late nineteenth-century Quebec, where a kind of Catholic fundamentalism had become widespread. It was associated with European ultramontanism and French-Canadian nationalism, and tended to be suspicious of making any accommodation with modern secular ideas, including modern science and medicine (as well as liberalism, democracy, and free speech). It was a strain of Catholicism not incompatible with deep suspicion of such "modern" practices as vaccination. In this view, epidemics of disease and their consequences were God's will. They were punishment for sin. The duty of the believer when faced with acts of God had to be acceptance and repentance.

Fatalism has always been the human default reaction to disease, suffering, and death. It is the reaction we have to events when we consider them inevitable—not when they are actually inevitable, but when they seem to be so. To a population that had always accepted the inevitability of deadly childhood diseases, expressed in the belief that the good God came from time to time and took away the little ones, a visitation of *picotte*, or smallpox, was simply what happened. The children were going to get it, and some were going to die. This was the way of the world, this was the way things had always been, and this was God's will.

The alternative of submitting to a strange new procedure advocated by the *maudit anglais*, which people said would only make the children more sick, was positively frightening. It was also frightening, in fact deeply angering, when authorities trundled sick children off to special smallpox hospitals, hotbeds of the disease, from which many of them never returned. It was invidious to have your home singled out for placarding and have to put yourself or the little ones in quarantine. How could you get food at the market or earn your wages? Some observers also noted the persistence among poor French-Canadians of the old folk practice of deliberately exposing infants to smallpox in the hope that they would get a small protective dosage.

You know, smallpox wasn't so bad, really. It didn't come that often. The death rate usually wasn't all that high, the disfiguration not all that bad. An infant's death in a very large family might not be as devastating as a similar death in a small family. The people feared vaccination more than they feared smallpox.

*

These were the ingredients of what became a toxic stew fueling resistance to vaccination, isolation, and quarantine among many of the poorer French of Montreal during the smallpox epidemic of 1885. The result was that the disease swept through the unprotected, passive, unresisting sections of the populace like flames through dry grass, an image Osler used in his textbook. Here was a major North American industrial city at the height of a century of enlightenment and progress, in which people succumbed to a completely preventable disease and died by the thousands, virtually every one of these deaths having been avoidable.

When the epidemic finally burnt itself out, in the winter of 1886, the more enlightened Montreal newspapers, supported by public opinion throughout North America, continued to fulminate about "the cruel murder of the infants of ignorant and misguided parents" that had taken place in Canada's largest city.[5] They urged an expansion of public health measures so that such ghastly events should never happen again in Montreal or Canada. In many churches, however, the dominant tone was to thank God for having lifted his chastisement on sinful Montrealers. In one

Protestant church, the minister drew attention to the "anthem of joy" in his denomination's service for the dead, in which the burdened soul exclaims, "I heard a voice from heaven saying unto me, from henceforth blessed are the dead which die in the Lord; even so saith the spirit; for they rest from their labors."[6]

The blessed children of Montreal who died from smallpox in 1885 still rest from their labor in Notre-Dame-des-Neiges and Mount Royal cemeteries. Their graves are testimonials to the survival in that era of a culture of fatalism, fear, and ignorance about fundamental health care procedures. Even as the epidemic took place, these attitudes were being superseded by a new culture of hope in and support for the efforts of modern healers to begin to make a difference.

THE SECULAR SAINTS
OF JOHNS HOPKINS

The first chapter began with the image of the young Dr. William Osler reading from the Bible and saying prayers at the bedside of a patient dying of smallpox, a disease Osler had no ability to treat. Fatalistically, Osler fell back on giving religious comfort to his patient, drawing on his upbringing as the son of an Anglican clergyman. Then I examined the deeper and much more tragic fatalism that accepted a terrible smallpox epidemic in Montreal in 1885 as an act of divine will, even though smallpox by then was fully preventable by vaccination.

To set this chapter in context, let us first consider the Osler family's evolution from parson to physician.

The Reverend Featherstone Osler, an ordained priest of the Church of England, was sent out to Canada in 1837 to be a missionary in backwoods settlements in the wilderness just north of the muddy town of Toronto. Battling enormous frontier adversity, Osler proved himself an outstanding success in his vocation. He founded churches and Sunday schools, ministered to the souls of growing congregations, and became greatly respected.

For a time, the Reverend Osler found that he was also looked to as a healer of patients' bodies. He became locally famous after a dark night when he was called out in the bitter cold to see a dying person:

I found the girl apparently very sick, and as they were expecting her to die many women were busy making her shroud. I found on examination that there was no sign of death, but they had persuaded her she was going to die, and she believed it. I ordered them to stop making the shroud, told her parents that I saw no immediate danger, prescribed a few simple remedies, made the girl take some nourishment, and left, promising to return in the evening. . . . I found the girl up sitting by the fire, and in a few days she was quite well, and was known for a long time afterwards in the locality as the resurrection girl.[1]

Parishioners asked Osler's advice on medicines to take and would not take medicines prescribed by their physician without his concurrence. If there had been a need to vaccinate any of his parishioners, Osler would have done it.

The Anglican circuit rider always believed that the core of his vocation was to preach the gospel of salvation through Jesus Christ, the wonderful consequence of salvation being eternal life and happiness. As regular physicians settled in his community, Osler phased out what he called his medical "practice." It was left to the youngest boy among his nine children—William, born in 1849—to become seriously interested in medicine, take training as a physician in Toronto and Montreal, and graduate from McGill University's Faculty of Medicine in 1872. After some eighteen months of postgraduate study in Britain and Europe, Dr. Osler was hired to teach the institutes, or fundamentals, of medicine, at McGill in 1874.

In the course of this chapter, William Osler will go on to far greater prominence than his father ever achieved as a backwoods clergyman *cum* physician. Before his fame, as a young unknown in Montreal, Osler not only saw private patients with smallpox, but also volunteered for extensive service on the smallpox ward of the Montreal General Hospital. Here in the 1870s he dealt with human suffering in one of its most ghastly, feared, and shunned manifestations, faced the uselessness of most forms of medication for the disease, and developed boundless admiration for the variety of women who gave nursing care to the sufferers.

*

Smallpox aside—and virtually everywhere except Montreal it was being put aside thanks to vaccination—the timing of Osler's medical vocation seemed almost providential. By the last third of the nineteenth century, the mainstream of regular medical practice was growing in respectability and promise everywhere in the Western world. Physicians were improving their training, their knowledge of the human body and its ailments, the scientific and empirical bases of their practice, and, in a few instances, their ability to treat pain and suffering.

By the 1880s, in large part because of the growth of pathological investigation, a competent physician stood a good chance of being able to accurately diagnose a disease condition. He could tell a patient what was wrong and then accurately predict the course of the illness. Doctors thus had a capacity for reassuring patients that was something akin to weather forecasting, what might be called medical meteorology. Note, therefore, how very useful physicians could be before they became curers. As well, the physician and his surgical colleagues now knew enough to deliver babies under conditions of cleanliness that saved many lives. They could anesthetize patients to save them from unbearable pain during childbirth and during surgery, and they could do many more operations, now working within the abdomen to remove, for example, diseased appendixes and gallbladders.

Nursing care—which had begun to flourish in part because of medicine's impotence and under the stimulus of sanitarianism—was becoming professional, effective, and increasingly important both inside and outside of hospitals. The use of opiates, especially morphine, made it possible to conquer many forms of pain. Public sanitation and vaccination were beginning to lift the menace of traditional plagues. Breakthroughs in what was beginning to be called medical research led to sensational advances, notably Louis Pasteur's discovery of a procedure for preventing rabies (which he called "vaccination" in homage to Jenner), and his and other investigators' discovery of a world of microbes, notably bacteria, that were the causes of many diseases. Empirical, scientific, research-based medicine seemed to hold out enormous promise. In the decade of the 1880s, wrote Osler's first biographer, "new discoveries were being announced like corn popping in a pan."[2] Confidence in the medical profession—which had been at a low ebb in the United States and Canada in the middle years of a skeptical, progressive century—was beginning to be restored.[3]

In North America by the late 1860s, a movement had developed to institutionalize higher standards of medicine by improving medical schools and upgrading hospitals. In the postbellum United States, the task seemed particularly urgent because of the earlier era around mid-century when anything and anyone had been allowed to flourish in health care. Standards of training and licensing had virtually disappeared in a democratic country where anyone was good enough to try to be a doctor. Quack and sectarian medicine had flourished. For example, the anti-vaccinationist "doctor" Alexander Milton Ross, who did so much harm in Montreal, called himself a physician by virtue of a few months' training at one of the United States' leading water-cure, or hydrotherapy, institutes. Other professional healers simply bought or invented their paper credentials.

The movement for educational reform in medicine began in Boston and gradually spread to other centers. It primarily involved turning medical schools, which had been mostly private proprietary enterprises run by physicians in their spare time, into graded, sequential, university-affiliated programs, bolstered by up-to-date teaching hospitals, libraries, pathology museums, and laboratories, and with a faculty of scientifically trained lecturers and distinguished clinicians.[4]

Montreal's McGill University, where young Osler first studied and then taught, had significantly less need of reform than many of the notorious medical diploma mills on the continent, such as Harvard's medical school. The Canadian college had retained more of the Old World standards that had been present from its founding than had most of its American counterparts, institutional offspring of a society that had violently rejected many things European. In the late 1870s and early 1880s, Osler not only taught at one of the best medical schools in North America, but as a hardworking, idealistic, personable, well-informed, well-traveled, research-interested, and usefully connected young physician, he seemed to personify medicine's promising future.

Thus Osler was judged a coming man by powerful members of the School of Medicine at the University of Pennsylvania and in 1884 was offered the chair of clinical medicine at America's oldest, and arguably most prestigious, medical school. Far more significantly, only five years later Osler was made head of medicine at America's youngest and most ambitious hospital and medical school, Johns Hopkins in Baltimore. He moved from an institution that had had a great past to one with a mandate to set the highest possible standards in the future.

FIGURE 8

Johns Hopkins Hospital, 1889. A temple of healing through scientific medicine.

Chesney Medical Archives, Johns Hopkins Medical Institutions

*

Johns Hopkins Hospital and University were the product of a great change in the direction of North American philanthropy. Throughout the nineteenth century, most giving by the wealthy had been religiously directed—to churches, seminaries, and other church-based charities. In 1867, however, the Baltimore merchant and financier Johns Hopkins—for reasons not totally known but certainly related to his being an adherent of the education-friendly Quaker sect—had drawn up a will bequeathing his then-vast fortune of $7 million (perhaps $300–$500 million in today's purchasing power) to found a university and a hospital mandated to compare favorably with like institutions anywhere in the world. The Johns Hopkins institutions, created faithfully by trustees after the merchant's death in 1873, were secular and scientific and uncompromising in their commitment to excellence. They trained professors and scientists and doctors, not preachers.

Johns Hopkins Hospital took several years to design and construct. It was not able to admit patients until 1889, and its medical school, a nominal partnership with Johns Hopkins University, did not open its doors until 1893 because of temporary shortfalls in endowment income. William Osler moved from Philadelphia to Baltimore in 1889 to become Physician in Chief of the hospital and would also serve as Professor of Medicine when the medical school opened. With no teaching to do in his early years at Johns Hopkins, Osler devoted much time to the writing of a textbook, *The Principles and Practice of Medicine*, published by D. Appleton and Company in 1892.

The opening of the Johns Hopkins Hospital, only four years after the ghastly Montreal events recounted in chapter 1, and then the opening of the School of Medicine another four years later were landmark events in the history of North American

FIGURE 9

Research training, Johns Hopkins. Salvation through science.

Chesney Archives, Johns Hopkins

medicine. Despite its relatively small size, Johns Hopkins was staffed by the best physicians and surgeons its trustees could find anywhere. It also had the highest admission standards of any American medical school, the best-equipped laboratories, ample funding, and probably the country's finest hospital facility, plus a knack for positive self-promotion through scholarly publishing and other activities. Johns Hopkins became the training ground for an elite group of medical scientists and practitioners, including nurses, who then fanned out to staff and lead hospitals and medical schools across the continent.

Notice in passing that the Baltimore journalist H. L. Mencken wrote in his memoirs about the lingering fear of Hopkins among poor blacks in Baltimore, the fear that if you or your body ever got into the hands of the Hopkins dissectors, they would cut you up like dead meat.[5] Even the most modern, best-staffed, and welcoming hospital North America could create still had to reckon with deep traditions of fear and fatalism. Gradually it overcame them. By 1910 and the publication of the famous Flexner Report to the Carnegie Foundation, *Medical Education in the United States and Canada*, the Hopkins medical institutions had become the model on which practically all of Flexner's recommendations for other centers were based.

The four founding medical fathers of Johns Hopkins became legends in their lifetime and in their specialties—William Welch in pathology, Howard Kelly in gynecological surgery and obstetrics, W. S. Halsted in general surgery, and, most significantly, William Osler in medicine. In his Johns Hopkins years, Osler became the best-known, best-loved, and most influential physician in America. He was the great American doctor, mentor, and role model. His influence would reverberate through the twentieth century and to the present in the latest round of activities and publications of the members of the American Osler Society, as well as in his ongoing influence at Johns Hopkins itself, still one of brightest beacons in North American and global medicine.

FIGURE 10

The Four Doctors by John Singer Sargent.
William Welch, W. S. Halsted, William Osler, and Howard Kelly—
founders of Johns Hopkins and North American medical excellence.
Chesney Archives, Johns Hopkins

What exactly was the Osler influence? Institutionally Osler helped usher in a vast improvement in American medical training by having senior medical students serve clinical clerkships on hospital wards. This was a step far beyond simple bedside teaching, in that it became for budding doctors a thorough introduction to work as real practitioners. Significantly, Osler had first experienced the clinical clerkship at McGill, which itself had adopted the practice from the medical school on which it had been modeled, that of the University of Edinburgh.

The second innovation was at the graduate level, as Osler, working with Halsted, expanded the fledgling intern system to create a category of resident physicians and surgeons who in a several years' training experience became highly qualified specialists. This system was also largely adapted from European, mostly German, practice. In both of these innovations, Osler, a Canadian, was helping to bring Old World state-of-the-art training to the United States.

A charismatic teacher and leader at Hopkins, who had literally written the book on medicine, Osler had immense influence on his students. Many of them went on to become outstanding physicians and medical educators. Osler was often hailed as the founding father of internal medicine in America, but his legacy is also claimed by neurologists, cardiologists, pediatricians, gastroenterologists, pathologists, and many others, not least family practitioners, whom he particularly respected. In fact, all of Osler's medical offspring were oriented toward empirically and scientifically based medicine. The principles advocated in *The Principles and Practice of Medicine* were observation, reliance on evidence, encouragement of research, ruthless self-criticism, and therapeutic open-mindedness. Indeed, so open-minded was Osler to whatever seemed to work best in attacking diseases that his most important protégé at Johns Hopkins was a physician who took up the specialty of surgery, a young Clevelander named Harvey Cushing.

*

Cushing, one of the fourth generation in his family to study medicine, went to Johns Hopkins in 1896 after graduation from Yale, Harvard Medical School, and internship at Massachusetts General Hospital. Physicians and surgeons worked together constantly on the wards at Hopkins, as many medical cases called for surgical intervention and vice versa. Osler himself was a particularly surgeon-friendly physician, not least because (as we see in more detail below) he was acutely aware of the limits of his specialties.

Medicine offered care, but few cures. By contrast, surgery in the late nineteenth century was making spectacular progress under the double impact of the introduction of anesthesia and asepsis, procedures that made it possible for surgeons to enter most cavities of the human body almost at will and with the prospect of doing their patients much more good than harm. Surgery worked. Surgery could eliminate disease. Surgery could make healing possible. From about the mid-1880s (appendectomies were first performed in 1886), surgery entered its first golden age. More than any other specialty, it became the driving force behind the transformation of hospitals from places for end-of-life care of the indigent into temples for the treatment of rich and poor alike.

Osler at Johns Hopkins appreciated surgery's possibilities, and in his textbook and practice did not hesitate to recommend resort to the knife where he thought it warranted. When young Cushing, for example, began attempting a surgical procedure to forestall intestinal perforation in certain cases of typhoid fever and questions were raised about his high mortality rate, Osler simply commented, "They all die with us."[6] When Cushing removed part of a diseased nerve from the thigh of a distinguished Johns Hopkins scientist, Simon Newcomb, the grateful patient hailed him as a miracle worker who had restored his ability to walk. "Prof. Newcomb believes himself to be entirely cured," a newspaper reported, "and leaves his crutches at Johns Hopkins as a souvenir just as the poor cripples who are cured by miracles leave theirs at Lourdes and at the shrine of St. Ann de Beaupre."[7] When even scientists were said to leave their crutches at Johns Hopkins, a new and powerful shrine was in the making.

Osler and his medical colleagues were powerless to treat the extreme facial pain caused by trigeminal neuralgia, or tic douloureux, a condition that drove many of its victims to madness or suicide. In 1899 Cushing operated on a patient who complained that in the early stages of a spasm he could feel "a devil twisting a red-hot corkscrew into the corner of the mouth," and that was just the start of the suffering. Cushing completely eliminated the patient's pain by severing the connection between the brain and the trigeminal nerve. When Cushing published details of his new procedure, the key to which was a better route to the nerve's junction with the brain, he reduced the mortality rate of these operations from approximately 20 percent to near zero. Osler commented that Cushing had "opened the book of surgery in a new place."[8]

When Cushing then began to experiment with the possibilities of doing surgery to relieve conditions affecting the brain itself, Osler, as head of medicine at Hopkins, gave the young man leeway to chart his own course, not forcing him to subordinate to hospital neurologists. There were many notable achievements in the early years of the Johns Hopkins Hospital—it had the resources and the ethos to supersede even the great German research institutes and universities—but none quite so spectacular as Harvey Cushing's development of successful neurosurgery in the first decade of the twentieth century. In a little operating room in Baltimore, Cushing became the first surgeon who could access the human brain at will and with the near certainty of doing more good than harm. He began to be able to relieve patients of torments that for many centuries had often been considered

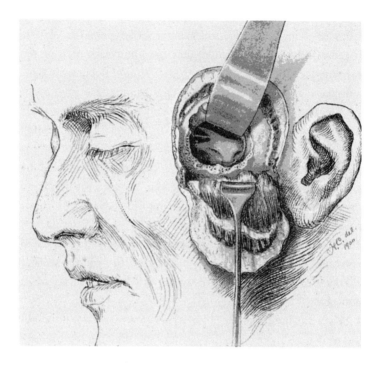

FIGURE 11

Harvey Cushing's ganglionectomy, 1900.

The surgical attack on tic doloureux.

Journal of the American Medical Association

the product of demonic possession. In some cases he could literally make the lame walk and the blind see. The second patient on whom Cushing performed a ganglionectomy for trigeminal neuralgia told his surgeon that he felt "resurrected."[9] At Johns Hopkins in 1908, the year Henry Ford introduced the Model T horseless carriage, Cushing removed a tumor from the brain of a completely conscious patient, talking with the man while he operated.

By 1910 Cushing had developed approaches to tumors located in the deepest recesses of the brain, those that caused pituitary disorders leading to severe distortions of growth, such as acromegaly and dwarfism. In passing, Cushing became the world's expert on the pituitary gland, previously a mystery organ, and began hypothesizing about the substances it apparently secreted and their effect on human development. His pituitary expertise made Cushing a leader not only in neurosurgery but also in the arcane and very young field of endocrinology, the study of the internal secretions of the body's glands.

As Cushing's fame grew, visitors from America and Europe began coming to his operating room in Baltimore to see the brain surgeon in action. When he moved to Boston in 1913 to become Surgeon in Chief of the showcase Peter Bent Brigham Hospital and Moseley Professor of Surgery at Harvard, foreigners flocked to see his operating techniques. Often the visitors would go on to see, first in Chicago and then in New York, Alexis Carrel, who could actually sew blood vessels together and was talking about transplanting organs and limbs, and who won the first Nobel Prize awarded for medical work done in America. They would go to Rochester, Minnesota, to study how the surgeons Charles and Will Mayo were turning their father's "clinic in a cornfield" into a hub for all kinds of modern medicine. They might stop in Cleveland to see the results of the effort that Cushing's friend George Crile was making to restore patients to life after their hearts had stopped beating, what he called "resurrecting" them.

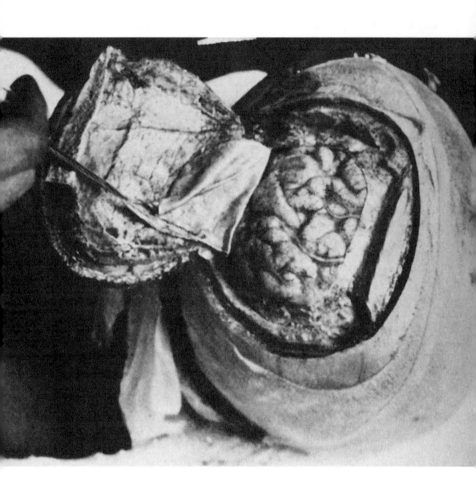

FIGURE 12

Cushing founds brain surgery, 1905.

Surgery, Gynecology and Obstetrics

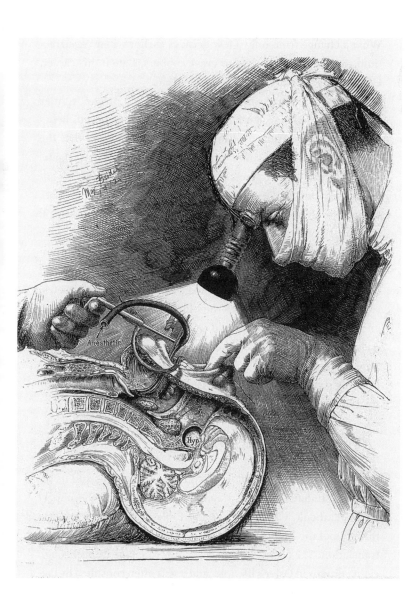

FIGURE 13

Cushing attacks pituitary tumors, 1912. Sketched by Max Brödel.

Brödel Archives, Art as Applied to Medicine, Johns Hopkins

What a change from only a few decades earlier when "resurrectionist" was a term applied to grave robbers. Late in his career, George Crile founded the Cleveland Clinic.

Surgery more than any other specialty stood on the frontier of Oslerian medicine. Fully conscious and very proud of being a medical frontiersman, Cushing commented in 1904 that the Northwest passage of surgery, which he was about to navigate, was the brain. The surgeons of course built on one another's advances. Cushing had learned a fair bit of his technique from W. S. Halsted at Hopkins, who is a towering figure in the technical history of North American surgery. But in every other way, as he was the first to admit, Cushing was deeply indebted to his mentor, neighbor, friend, and role model, the physician Osler. (Not the least of the reasons why Halsted could not serve as a professional guide to Cushing was the older man's struggle to break an addiction he had developed in his surgical youth to a local anesthetic, cocaine. Halsted had managed to get off cocaine, but only by becoming secretly addicted to morphine instead. During Cushing's years as his resident, Halsted's professional and personal behavior was, to be charitable, often eccentric.)[10]

*

Osler's influence extended far beyond Johns Hopkins and its influential students. It was spread most obviously by the impact of his textbook, which became the Bible of medical education throughout the English-speaking world and in other countries in translation. *The Principles and Practice of Medicine* went through eight editions in Osler's lifetime, and it was often consulted by laymen as well as physicians and students, and so helped to spread the gospel of the new scientific medicine around the world.

The most famous and most important lay reading of Osler's text was in the summer of 1897 by an American Baptist minister, the Reverend Frederick Gates. Gates happened to be the philanthropic adviser to the fabulously rich Baptist oil magnate John D. Rockefeller. Most of Rockefeller's substantial philanthropies had been to causes supported by the Baptist Church. Gates — fascinated by both the achievements of recent medical research and the challenges medicine faced, as outlined by Osler — decided to recommend that Rockefeller begin supporting medical research in a major way. In 1903 the Rockefeller Institute for Medical Research was established in New York (Alexis Carrel spent most of his career there), and within a few years Rockefeller money was supporting a wide variety of initiatives in health care and medical education. An up-to-date philanthropist now gave at least as much money to health causes as to religious ones.

Worried that he was burning out, Osler left Johns Hopkins in 1905, eight years before Cushing did, to take up a sinecure as Regius Professor of Medicine at Oxford, where he spent the rest of his career. He was of course much honored. In 1911 he was awarded a baronetcy, thus becoming *Sir* William. When a little girl who had been a patient of his heard the news, she exclaimed, "Oh dear. They should have made him King."[11] Perhaps that was an exaggeration, but Osler had played a critical role in the building of Johns Hopkins as the *sanctum sanctorum* of healing in America, for which he was elevated into something like the patron saint of both Hopkins and the advent of modern medicine. In a famous Max Brödel cartoon, Osler hovers angelically over Johns Hopkins, as the bacteria flee. "We all worship him" was a comment endorsed by medical men and women on the staff of Johns Hopkins.[12]

I am using religious metaphors deliberately, for to Osler the language was more than metaphorical. Osler was very much his father's son in having a sense of his profession as a vocation, a calling, to which he was as dedicated as a priest in the service of religion. He thought of medicine as service to humanity, and he thought of it as service of a high order, what he called "our ministry of health."[13] Harvey Cushing was in most other ways

FIGURE 14

Osler by Max Brödel, 1896. Microbes flee the saint of Johns Hopkins.

Chesney Archives, Johns Hopkins

temperamentally different from Osler, but felt exactly the same as Osler did about medicine. Both men agreed with Stephen Paget, writing in *Confessio Medici*: "If a doctor's life may not be a divine vocation, then no life is a vocation, and nothing is divine."[14]

This was not a superficial or transient identification, especially for Osler. In the fullness of his career, he gave many addresses to doctors and medical students on the state of the profession, lay sermons, in which he elaborated at great length on why modern scientific medicine should be understood as "the Promethean gift of the century" to humankind, the gift that made possible "man's redemption of man."[15] Osler argued that medicine, a catholic and borderless brotherhood, had done more than any other profession, including the church, to alleviate the ills of suffering humanity. In one of his 1905 farewell addresses on leaving America, Osler spelled out his estimate of the contribution made by his profession in his lifetime:

Medicine is the only world-wide profession, following everywhere the same methods, actuated by the same ambitions, and pursuing the same ends. . . . Similar in its high aims and in the devotion of its officers, the Christian Church . . . yet lacks that catholicity . . . which enables the physician to practise the same art amid the same surroundings in every country of the earth. . . . In a little more than a century a united profession, working in many lands, has done more for the race than has ever before been accomplished by any other body of men. So great have been these gifts that we have almost lost our appreciation of them. Vaccination, sanitation, anaesthesia, antiseptic surgery, the new science of bacteriology, and the new art in therapeutics have effected a revolution in our civilisation . . . a revolution which for the first time in the history of poor, suffering humanity brings us appreciably closer to that promised day when the former things should pass away, when there should be no more unnecessary death, when sorrow and crying should be no more, and there should not be any more pain.[16]

In other addresses, Osler wrote lyrically of mankind imprisoned and helpless for centuries in the shadow of death, "singing . . . vain hymns of hope, and praying vain prayers of patience" until the advent of modern science began unlocking the mysteries of nature and finding "a way of physical salvation."[17]

*

Osler's was a powerful secular humanist vision, but it was wildly optimistic. As he always realized, it was deeply vulnerable to complaints about the inability of physicians, even at the dawn of the twentieth century, actually to cure. Doctors might be able to prevent the development of smallpox, cholera, and other infections, but they could not cure anyone who became afflicted. They now knew what caused tuberculosis, but they could not cure it. They could not cure pneumonia; they could not cure cancer; they could not cure diabetes—they could not cure much of anything. They could prevent, diagnose, and predict, and they could alleviate pain. But very many of the medicines and procedures they had inherited from past generations were actually useless or even harmful—as Osler well knew and taught, giving him a reputation as an advocate of pharmacological therapeutic nihilism. He understood that often the best aid doctors could offer was to fall back on pantheistic hope in the curing power of nature and time.

Very often, especially when it came to dealing with what appeared to be functional disorders, perhaps psychically generated, physicians seemed less effective than even faith healers, the religious leaders who taught the virtues of appealing to the deity or the saints for intercession, say, by making pilgrimages to famous shrines. In Osler's America and among the class from which his own private practice was largely drawn, there was a great vogue for the teaching of Mary Baker Eddy, high priestess of the sect known as Christian Science, to the effect that the Bible revealed the unreality of all disease.

At its best, Osler argued, medicine did cure and it would cure more. In the gray area where mind and body interacted to create paralysis, hysteria, neurasthenia, and like disorders, if patients had as much faith in their physicians as they did in their clergy, if they had faith in Saint Johns Hopkins Hospital, they would experience the same cures that the faith healers offered. "I have had cases any one of which . . . could have been worthy of a shrine or made the germ of a pilgrimage," Osler wrote in his essay "The Faith that Heals," and went on to give an instance:

For more than ten years a girl lay paralysed in a New Jersey town. A devoted mother and loving sisters had worn out lives in her service. She had never been out of bed unless when lifted by one of her physicians. . . . The new surroundings of a hospital, the positive assurance that she could get well with a few simple measures sufficed, and within a fortnight she walked round the hospital square. This is a type of modern miracle that makes one appreciate how readily well meaning people may be deceived as to the true nature of the cure effected at the shrine of a saint. Who could deny the miracle? And miracle it was, but not brought about by any supernatural means.[18]

Osler pointedly suggested that the devout would be wise to hand over the problem of mind-body cures to the medical profession. "The less the clergy have to do with the bodily complaints of neurasthenic and hysterical persons the better for their peace of mind and for the reputation of the Cloth."[19]

The demand that Osler and his generation of physicians made on patients was that they now trust their doctors. Faith had to be transferred to the physician because he could diagnose, predict, and, in some cases, heal. You had to believe in your doctor. Patient testimony was almost universal to the effect that a doctor like Osler commanded trust and loyalty. Edith Gittings

Reid, a writer whose family members were all treated by Osler, described his charismatic bedside manner:[20]

To have been a patient of Sir William Osler's . . . was to have obtained an almost impossible idea of what a physician could be. . . . It was not necessary for him to be sensitive to a social atmosphere, because he always made his own atmosphere. In a room full of discordant elements he entered and saw only his patient and only his patient's greatest need, and instantly the atmosphere was charged with kindly vitality, everyone felt that the situation was under control, and all were attention. No circumlocution, no meandering. The moment Sir William gave you was yours. . . . With the easy sweep of a great artist's line, beginning in your necessity and ending in your necessity, the precious moment was yours, becoming wholly and entirely a part of the fabric of your life. . . .

Such telling love, such perfect confidence were given him that he could do what he liked without causing offence.*

The great triumphs to come, in Osler's view, would involve medicine actually being able to cure organic disorders. Having

---

*Osler was particularly noted for his ability to fascinate children. A British colleague described Osler at Oxford as once missing a prognosis, but getting fine results by understanding the psychological impact of presenting himself as a doctor *cum* magician: "One remembers a young brother with very severe whooping-cough and bronchitis, unable to eat and wholly irresponsive to the blandishments of parents and devoted nurses alike. Clinically it was not an abstruse case, but weapons were few and recovery seemed unlikely. The Regius, about to present for degrees and hard pressed for time, arrived already wearing his doctor's robes [gowns]. To a small child this was the advent of a doctor, if doctor it in fact was, from quite a different planet. It was more probably Father Christmas.

"After a very brief examination this unusual visitor sat down, peeled a peach, sugared it and cut it in pieces. He then presented it bit by bit with a fork to the entranced patient, telling him to eat it up, and that he would not be sick but would find it did him good as it was a most special fruit. Such proved to be the case. As he hurried off Osler, *most uncharacteristically*, patted my father on the back and said with deep concern 'I'm sorry, Ernest, but I don't think I shall see the boy again, there's very little chance when they're as bad as that.' Happily events turned out

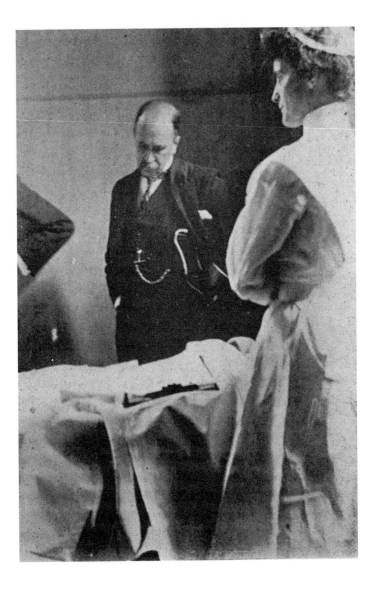

FIGURE 15

Osler at the bedside. A trusted physician with limited power.

Osler Library, McGill University

witnessed the dawn of bacteriology, he could see, for example, that science would one day discover antibacterial agents that would cure tuberculosis or cholera or pneumonia. In the 1890s he experimented hopefully with the "tuberculin" that the famous German bacteriologist Robert Koch offered as a cure for consumption. It failed. But a therapy that did not fail in the 1890s was the feeding of thyroid extract to patients suffering from cretinism and myxedema, disorders caused by thyroid deficiency. The treatment had emerged from experimental work in several centers. Osler himself used thyroid extract at Hopkins and made a special study of his profession's experience treating cretinism. He asked colleagues to send him before-and-after pictures of their patients, and he published a number of these in an article he wrote in 1897 on sporadic cretinism in America.

Even in scientific articles Osler wrote rich prose, full of allusions:

No type of human transformation is more distressing to look at than an aggravated case of cretinism. It recalls Milton's description of the Shape at the Gates:

"If shape it might be called, that shape had none
Distinguishable in member, joint or limb."

---

otherwise, and for the next forty days this constantly busy man came to see the child, and for each of these forty days he put on his doctor's robes in the hall before going in to the sick room.

"After some two or three days, recovery began to be obvious and the small boy always ate or drank and retained some nourishment which Osler gave him with his own hands. If the value of personal approach, the quick turning to effect of an accidental psychological advantage (in this case decor), the consideration and extra trouble required to meet the needs of an individual patient, were ever well illustrated, here it was in its finest flower. It would, I submit, be impossible to find a fairer example of healing as an art." Patrick Mallam, "Billy O," in *Oxford Medicine,* ed. Kenneth Dewhurst (London: Sandford, 1970), 94–99.

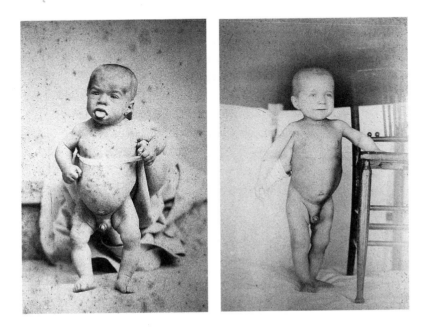

FIGURE 16A, B

Cretinism treated by thyroid extract, 1897. The first miracle treatment.

American Journal of the Medical Sciences

[O]r those hideous transformations of the fairy prince into some frightful monster. The stunted stature, the semi-bestial aspect, the blubber lips, *retroussé* nose sunken at the root, the wide-open mouth, the lolling tongue, the small eyes, half closed with swollen lids; the stolid, expressionless face, the squat figure, the muddy, dry skin, combine to make the picture of what has been well termed the "pariah of nature."

Not the magic wand of Prospero or the brave kiss of the daughter of Hippocrates ever effected such a change as that which we are now enabled to make in these unfortunate victims, doomed heretofore to live in hopeless imbecility, an unspeakable affliction to their parents and to their relatives. . . . The pictures . . . emphasize as words cannot the magical transformation which follows treatment.[21]

Here was the beginning of the dream come true.

∗

William Osler was a one-man hypotenuse of the North Atlantic medical triangle. As a physician shaped by and shaping British, Canadian, and American cultures, his life and career serve as an almost perfect example of how this triangular metaphor, no longer used in the study of politics or foreign affairs, can still nicely illuminate a cultural and scientific transition. Osler brought the best of Old World medicine to the United States through Canada. In Osler's hands, by patients' bedsides at the well-endowed Johns Hopkins Hospital in Baltimore, it was very good medicine indeed. In fact, as the direction of transatlantic medical pilgrimages began to reverse, North American medicine was metamorphosing from colonial subordination to Europe into a new role of global leadership. Sometimes on the individual level, medicine even wrought magical transformations.

William Osler never forgot his Canadian roots, and, as we see in the next chapter, Canada never forgot Osler. The great American physician effectively paid no attention to the Canada-U.S. border—or any other border for that matter, as he loathed nationalism and chauvinism in medicine; he and Cushing and their disciples were truly doctors without borders before that term was brilliantly appropriated by a humanitarian organization. As we see in more detail in chapter 3, Osler was a constant presence back in Canada. For now it is enough to notice a major address he gave in 1909 to the Ontario Medical Association, meeting in Toronto, entitled "The Treatment of Disease." In it he developed many of his familiar arguments about the wonderful progress of medicine in the nineteenth century, with references to "the Napoleonic campaigns which medicine is waging . . . glorious days. . . . Nothing has been seen like it on this old earth since the destroying angel stayed his hand on the threshing-floor of Araunah the Jebusite."[22] No one could cite scripture to illuminate the progress of medicine better than Osler, the clergyman's son, still reflecting the influence of his upbringing.

On a secular note, in that 1909 address Osler instanced the use of thyroid extract as a metabolic treatment compensating for thyroid insufficiency. He told his Toronto audience that other conditions resulting from the perversions of metabolic functions, such as Addison's disease and acromegaly (the latter a clear reference to Cushing's very recent work), would soon be treatable. Medical science was beginning to unravel the mysteries of the secretions of hormones from glands, raising the prospect of elevating organotherapy from quackery to serious therapy. Then Osler added, prophetically, "and as our knowledge of the pancreatic function and carbo-hydrate metabolism becomes more accurate we shall probably be able to place the treatment of diabetes on a sure foundation."[23]

III

MASTERY: TORONTO, 1922

The medical border between Canada and the United States was more open in William Osler's time than it is in today's era of differing systems of health insurance. In the nineteenth century, Canada's leading urban centers, Montreal and Toronto, saw themselves as vying for economic, educational, and social leadership with the great cities of the eastern and central United States. This was particularly true in health care. Canadians aspired then, as they do now, to have medical institutions as advanced as any on the continent. I have mentioned, for example, the high quality of McGill University's Faculty of Medicine, both as it shaped the young Dr. Osler and then was shaped by him. After Osler left McGill for the United States, its faculty struggled self-consciously for decades to continue to adhere to the highest Oslerian standards, and in the long run it has largely succeeded. Throughout its history, McGill's has been one of the premier medical schools in continental North America.

But Montreal after the creation of the Dominion of Canada in 1867 found itself struggling to maintain economic hegemony in the face of the rise of the Ontario economy, dominated by Toronto. Similarly, McGill University's educational leadership gradually began to be challenged by the University of Toronto. The early history of medical instruction in Ontario is dauntingly complex and more than a little chaotic. As young Osler found at the end of the 1860s when he took two years of training at the proprietary Toronto School of Medicine, the story is not exactly a study in competence. By the 1880s that situation had changed, as the settled and prospering agricultural regions of southern and central Ontario began to produce crops of eager, talented young men—such as the Osler boys—who aspired to the best educational and professional opportunities North America could provide. If they could not find these opportunities at home in Canada, they would of course look south to the United States.

In the 1880s the University of Toronto began to develop its Faculty of Medicine, with particular emphasis on strengthening the scientific underpinnings of medical training. Under the leadership of men like Daniel Wilson, Ramsay Wright, William Aikins, and A. B. Macallum, Toronto created what soon became one of the largest and best medical schools on the continent. Osler himself, who never forgot his Ontario origins, had foreseen and tried to stimulate such development, writing as early as 1881 about his dream of a model hospital and medical school in "Otnorot," institutions so good "that finally one could study better at Otnorot than in cultured and civilized Europe."[1] While that possibility was gradually unfolding, many of the impatient, energetic young sons of Ontario followed the paths Osler was beating, first to McGill and then to Johns Hopkins and other top-notch American universities.

The Canadian involvement at Johns Hopkins, what Sandra

McRae has called the "Canadian Club" there,[2] was particularly noticeable and important. Well before Osler, there were several young Canadians, including Ramsay Wright and A. B. Macallum, who studied at Johns Hopkins University and then returned to build science education at the University of Toronto. Osler's heyday in the 1890s saw heavy traffic on the medical freeway connecting Canada with Maryland. So many of William Osler's residents, students, and coworkers wore Canadian fur caps that he became worried about being accused of nepotism. When once asked at a dinner where he came from, he replied, "Far Northern New York."[3] The successors of three of the four founding fathers of Johns Hopkins were Canadians, and a strong Canadian presence continues to the present at Hopkins.

As we set the stage for the coming of globally significant medical events in Canada, it is not irrelevant to consider a possible explanation for Canadian strength in medicine and health care. This may involve the fact of English-Canadian society having always had a conservative tilt, compared with that of the United States. Shaped by Great Britain, and evolving rather than rebelling against the mother country, Canada was a place where the traditional professions always maintained a higher social status than they did in the seething democratic republic to the south. The difference may still linger today, just as it may also be true that Canadians are not as drawn to purely business and entrepreneurial pursuits as much as Americans. The other side of that coin might be that Canadians' Americanness serves to energize their work as professionals, fostering a drive to high levels of performance. It was said of the Osler boys, by one of their English relatives, that they were particularly fine specimens, being "English gentlemen with American energy."[4]

At their best Canadians were never content to be colonials. By the beginning of the twentieth century, the University of Toronto, like McGill, aspired to rival rather than complement Johns Hopkins as a center for the best medical work being done on the continent—or anywhere else. In 1903 Toronto opened a new state-of-the-art medical building to house a unified and growing medical school. Of course, William Osler was invited to come up from Hopkins to give the celebratory address at the opening ceremony. In that speech,[5] Osler talked of his hope that the University of Toronto's Faculty of Medicine would become "a school of the first rank in the world."*

As a doctor whose graduate work had been financed by loans from a rich brother, and as a founder of Johns Hopkins, Osler understood the importance of persuading philanthropists to give their money to help build great medical institutions. He urged medical educators to follow the lessons of the clergy in seeking endowments from wealthy benefactors. In Toronto in the early twentieth century, there was no more energetic benefactor than another son of the Ontario countryside, an Irish-Canadian from Peterborough named Joseph Wesley Flavelle, who in the 1890s made a great fortune exporting the bounty of Ontario farms to Britain in the form of sides of premium bacon (and in passing was responsible for Toronto receiving its enduring nickname, "Hogtown"). A devout Methodist, Flavelle had from early manhood taken an interest in charitable hospital work. In the early 1900s as chairman of the board of the Toronto General Hospital, he undertook a major modernization and transformation of the city's major, venerable, but often neglected house of charity.

Under Flavelle's leadership, the old Toronto General Hospital was entirely rebuilt in the first decade of the twentieth century at a new location near the university and with an important new role as the university's primary teaching hospital. Flavelle and

---

*He went on famously to address the question of "the master-word in medicine": Osler declared that the master word is "work," and his address became an inspiration to generations of ambitious medical students, most of whom in 1903 were still too often tempted to waste their energies in idleness, drink, and other dissipations. Perhaps we can take today's medical students' work ethic for granted and change the master word to something like "balance."

General Hospital, Toronto, Canada

FIGURE 17A, B

Medical building, University of Toronto/
Toronto General Hospital, 1912. Fields of medical dreams.

Fisher Library, University of Toronto

other public-spirited citizens, including department store magnate Timothy Eaton, envisioned Toronto General as the commanding hospital in a great Canadian city. When he and his colleagues sought organizational models for their new institution, aiming to emulate excellence, they naturally turned to William Osler's handiwork at Johns Hopkins.

Simultaneously—these Canadian equivalents of Carnegie and Rockefeller were men of great energy—Flavelle was chairman of an Ontario Royal Commission charting the future of the University of Toronto. Its 1906 report became a landmark in the modernization and transformation of that institution. Ontario's premier university—which aspired to be Canada's premier university and one of North America's leaders—would now be more efficiently governed, receive more resources from the province, and develop its graduate training programs and its capacity to do pure and applied research. An integral part of that new research thrust would rest on partnership between the Faculty of Medicine and the Toronto General Hospital, not unlike the partnership between the Johns Hopkins medical institutions in Baltimore.[6]

Any doubt about who and what was inspiring the Torontonians' vision would have been forever ended if the effort by Flavelle and his friends to persuade William Osler to become president of the University in Toronto in 1906 had succeeded. It did not. By 1906 Osler was settled in semi-retirement in Oxford. But he did take the offer seriously, and for the rest of his life he continued to be an ardent supporter of Canada's great hospitals and medical schools.

The Faculty of Medicine at Toronto, along with McGill's, was given top marks in the Flexner Report of 1910, its laboratories singled out as comparable to the world's best. With the opening of the new Toronto General Hospital in 1913, the university's clinical resources were also second to none in North America. The good research labs were now being used by advanced graduate students, such as James B. Collip, a Belleville, Ontario, na-

tive who had taken an elite honors degree in physiology and biochemistry at Toronto in 1912 and had decided to bypass medical
school in favor of a career as a researcher. In 1916 Collip became
one of the university's first doctors of philosophy when he was
awarded a Ph.D. degree in biochemistry; he took his first job as
a founding member of the faculty of the medical school at the
University of Alberta. Also that year a young Toronto practitioner, Walter Campbell, found that the city had the capacity to
enable him to begin to specialize in a disease that deeply interested him; he became head of a unique diabetes ward at the Toronto General, working under the highly regarded Professor of
Medicine Duncan Graham. Continuing its thrust to improve its
staff, in 1918 the University of Toronto succeeded in luring Professor James J. R. Macleod away from one of the United States'
great universities, Western Reserve in Cleveland, to assume the
chair of physiology.

At the end of the Great War, news that Toronto and McGill
would receive very large benefactions from the Eaton family and
from the Rockefeller Foundation—the latter's interest in health
having been inspired by Osler—made it seem likely that Canada's major medical schools would continue to be important players in the postwar expansion of North American medicine. To
undertake ongoing research in problems involving carbohydrate
metabolism, Toronto's J. J. R. Macleod, for example, was splen-
didly supported. He had access to the university's research animal facility, funds to hire student assistants, and even an animal
operating room in the medical building so advanced that it had
been sitting unused by anyone for the facility's first fifteen years.
Macleod was so satisfied with his situation at Toronto that in
1920 he declined to enter his name as a candidate for the chair in
physiology at Johns Hopkins.

There was every prospect of further growth in his department's
research capacity, including serious negotiations with that promising biochemist Collip, who was finding teaching at the fledgling

University of Alberta a bit confining. Pending a possible move, Collip arranged to spend considerable time in Toronto in 1921–22 doing research with Macleod, financed by a Rockefeller Travelling Fellowship. For philanthropic purposes, the Rockefellers ignored the U.S.-Canada border. Collip was particularly impressed by the high quality of the research opportunities available to him in Toronto.[7]

*

The research ideal was so pervasive at Toronto that even ordinary undergraduate medical students were encouraged to believe their future could include contributing to new knowledge. Thus Fred Banting—a farm boy from Alliston, Ontario, close to William Osler's birthplace—was interested in research despite an undistinguished academic record. Born in 1891, Fred had dashed his parents' hope that he would become a Christian minister and instead had taken his medical degree at Toronto in 1917. After service as a medic in France, Fred began to take specialized training in the hottest area of medicine, surgery, at Toronto's Hospital for Sick Children, and became involved in several projects with a research component. His hospital appointment was not renewed, so he left Toronto in 1920 to begin general practice in London, Ontario, but he still toyed with the idea of making a research contribution. The toying became more active tinkering when he found he had few patients to fill his time.

He also filled time and earned a few dollars as a demonstrator in surgery and anatomy at the University of Western Ontario's small and as yet not very research-oriented medical school. It was in his capacity as demonstrator, charged with having to talk to medical students about the pancreas, that Banting read an article in the leading surgical journal *Surgery, Gynecology and Obstetrics* that stimulated his thinking and caused him sometime in the early hours of October 31, 1920, to jot down an idea for a research

FIGURE 18

Dr. Frederick Banting, born in Osler's neighborhood.

Fisher Library, University of Toronto

project on the pancreas that might possibly lead to a treatment of "diabetus," a disease with which Banting had no familiarity or experience:

DIABETUS

Ligate pancreatic ducts of dog. Keep dogs alive till acini degenerate leaving Islets.

Try to isolate the internal secretion of these to relieve glycosurea.[8]

*

Diabetes mellitus had been recognized for millennia as a melting down of the flesh into urine, what one practitioner had called "the pissing evile."[9] The modern era of diabetes knowledge began in 1889 with the accidental discovery by a European research physiologist, Oskar Minkowski, that in its most severe manifestation, now called type 1 diabetes, the disorder is generated by a problem with the pancreas, not, as had previously been believed, the liver or the stomach. Minkowski's hypothesis that there must be a mysterious something in the pancreas that regulated metabolism was then given increased credibility by other researchers' gradual discovery of powerful secretions from a number of glands, including the thyroid. Abnormalities in the flow of these internal or endocrine secretions appeared to be the cause of a number of disorders. As we saw in chapter 2, the wonderful discovery that cretinism and myxedema could be cured by administration of thyroid extract seemed to hold out the promise of treating other diseases, including diabetes, with similar organotherapeutic methods—thus Osler's 1909 prediction in Toronto that the treatment of diabetes would eventually be put on a sure footing.

Nevertheless, when Fred Banting wrote down his idea in the autumn of 1920, some thirty years after Minkowski's discovery sparked substantial research into the pancreas and diabetes,

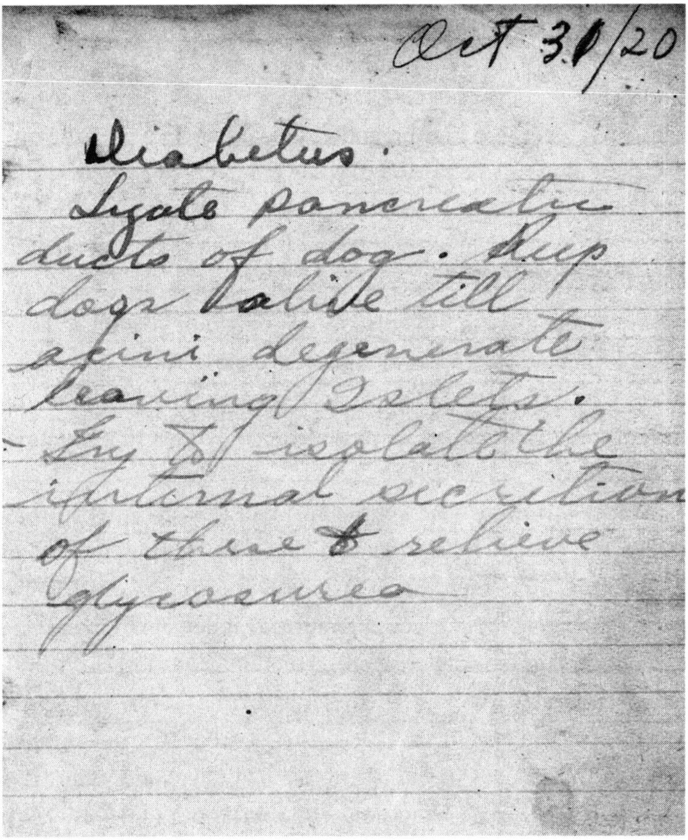

FIGURE 19

Banting's idea for the research project that
culminated in the discovery of insulin.

Fisher Library, University of Toronto

there had still been very little progress. Juvenile, or type 1, diabetes was a horrible disease. Like smallpox, it could not be treated medically. The metabolic failure involved in diabetes could not be reversed. The chain of consequences for the body of metabolic breakdown—hunger, thirst, constant urination, weight loss, coma—seemed inexorable. When diabetic patients became comatose, there was nothing more to do other than alert the undertaker, pray, or ask a priest to come to administer last rites.

At best, the ravages of diabetes could be delayed by systematic lightening of its metabolic burden, in other words, having people with diabetes eat less. Thus by 1920 the state-of-the-art therapy for the type 1 diabetic was the "under-nutrition" regimen advocated by the world's leading diabetes researcher, Dr. Frederick Allen, who had carried out years of experiments at Harvard and the Rockefeller Institute, all of which came down to restricting calorie intake.

"Under-nutrition" was a euphemism for starvation. The progress of the "successfully" treated diabetic was to prolong life for a few months or a few years by slowly starving to death. Few patients or their parents could endure such denial of our fundamental need for food. There were exceptions: the most successful patient we know of on an Allen diet was James Havens of Rochester, New York, who lasted for six years, from 1915 to 1921, before becoming bedridden and near death at age twenty-one. "His condition is not such as to hold out any hope," James Haven's father wrote, fatalistically, in 1921.[10] In the spring of 1922, young Havens's hospital chart recorded that he was crying much of the time and was "anxious to die and end his misery."[11] We do not know whether the Havens family sought the comforts of faith.

The fact that Osler's 1909 prediction still had not come true more than a decade later was a depressing reflection on an enormous amount of medical research that had proven fruitless—quite apart from the ways in which medicine and science had just been diverted into becoming instruments of death and destruction during the Great War, an occurrence that Osler himself found profoundly depressing. Over a thirty-year period, several hundreds of attempts around the world to find some pancreatic substance that could restore metabolism had failed. All the splendid universities and hospital wards and research labs, even the resources of the Rockefeller Institute where Frederick Allen did his research, had not been able to produce a diabetes breakthrough. Allen's stop-gap therapy for diabetes was disappointing, almost impossible for normal human patients to follow. By 1919 Allen had left the Rockefeller Institute, and it had ended its initiative in diabetes (incidentally putting out of employment a bright researcher, Israel Kleiner, who published a paper that year on his work with pancreatic extracts, which if anyone had read it carefully and had the resources to do follow-up studies could have led to the discovery of insulin. Kleiner himself had to take a job at the University of Kansas, which did not have the resources to do animal research, and he stopped his work on diabetes.[12])

In fact, by 1919 the whole promise of the young discipline of endocrinology seemed a cruel disappointment. The triumph of thyroid replacement therapy in the 1890s was not being replicated. In Boston even Osler's great protégé, Harvey Cushing, was becoming disillusioned with his pituitary work, as few of his hypotheses about the gland's action and few of his therapies for pituitary disorders seemed useful. Cushing and others wondered if the whole idea of endocrinology was flawed. Perhaps the field was little more than a stew of old wives' tales and folklore about vital juices and essences. Cushing sometimes called it "endo-criminology." To his surprise and chagrin, Cushing was elected president for 1920–21 of the four-year-old Association for the Study of Internal Secretions, today the very large Endocrine Society. His presidential address was mostly an attack on the quackery and unrealizable expectations promoted by the society's founders.[13]

By this time a number of diabetes researchers wondered if

Minkowski's work had put everyone on the wrong track. Perhaps there was no such thing as an internal secretion of the pancreas. Certainly nothing very therapeutically useful was coming from all the work being done on diabetes and the study of other internal secretions. "Personally I have not known an instance of recovery in a child," Osler wrote of diabetes in the 1918 edition of *The Principles and Practice of Medicine,* nine years after he had predicted its treatment would soon be on a sure foundation.[14] Even most smallpox patients had recovered way back in the 1870s and 1880s. On January 18, 1922, Harvey Cushing wrote to a correspondent that "the disorders of the thyroid are about the only ones, which, so far as we know, can be very successfully benefitted by glandular administration."[15] So much for the promise of modern scientific medicine.

*

Wonderful and unexpected irony. In a laboratory at the University of Toronto on January 19, 1922, the day after Cushing wrote that letter and of course totally unbeknownst to him, J. B. Collip discovered a method of extracting from Fred Banting and Charley Best's crude pancreatic substances an active principle that appeared to work in the treatment of diabetic animals. On January 23, on the diabetes ward of the Toronto General Hospital, the substance was given to a diabetic human child, Leonard Thompson, with spectacular results. University of Toronto researchers now realized that they were making one of the great breakthroughs in the history of modern medicine.

*

The research process that began with Banting's idea and ended in Collip's production of an effective pancreatic extract has be-

come a fairly familiar story since the 1982 publication of my book
The Discovery of Insulin swept away a number of myths and mis-
conceptions. There is no real doubt that Banting and his student
assistant Best would not have reached insulin without the advice of Macleod and the expertise of Collip. Fred Banting was
an interesting person, clever and insistent, but he had no training in research, a very crude understanding of diabetes and the
pancreas, and even rather limited surgical experience (he had
great trouble doing total pancreatectomies and preventing infection in his animals). Charles Best was a student helper, no more,
no less.

It is virtually certain that nothing would have come of Banting's idea had he not been advised at the University of Western Ontario to go to the University of Toronto, where he could
draw on J. J. R. Macleod's expertise and resources. Thus, almost
serendipitously, an eager young graduate with an interesting research idea appeared on Macleod's doorstep at a time when Macleod had unused research capacity—extra dogs, extra students,
an unused animal operating room, even an expert biochemist in
the form of Collip, who had spare time on his hands and could
be brought into the research when it began to look promising.
More generally, Banting's idea, which in fact was physiologically unsound (the cell types in a duct-ligated pancreas do not
atrophy differentially), could not have been pursued with even a
faint hope of useful results before the twentieth century's second
decade: only then had other researchers developed techniques
permitting the relatively quick accumulation of serial blood
sugar readings, necessary to track the effects of the experimental
administration of pancreatic substances. To use a military comparison, the Toronto breakthrough had come only after a massive
buildup of resources and prolonged battering at the enemy along
many fronts. Four hundred previous assaults on the strong-
hold of diabetes had failed, Macleod estimated. The well-armed,

DR. J. B. COLLIP   DR. J. J. R. MACLEOD   DR. F. G. BANTING   DR. C. H. BEST

Photographs by M Lynnde (Toronto)

Doctors Macleod and Banting were jointly awarded the $40,000 Nobel prize for the discovery of insulin. Dr. Macleod promptly divided his share with Dr. Collip of Alberta University; while Dr. Banting telegraphed to his colleague, Dr. Best, "You are with me in my share always."

THE FOUR CANADIAN SCIENTISTS WHO SHARE THE GLORY AND THE PRIZE FOR DISCOVERING INSULIN

INSULIN WINS A GREAT PRIZE

FIGURE 20

The discovers of insulin: J. B. Collip, J. J. R. Macleod, F. G. Banting,
and C. H. Best. Uneasy collaborators, covered in research glory.

Department of Biochemistry, University of Western Ontario

expert, multi-stage Toronto attack, exploiting a lucky initial probe, succeeded.

The success—the payoff—was immediate and spectacular, mirroring the dramatic triumphs of thyroid therapy a quarter century earlier but affecting a much larger population. Fourteen-year-old Leonard Thompson was given another thirteen years of life on insulin. The daughter of U.S. secretary of state Charles Evans Hughes, fifteen-year-old Elizabeth Hughes, had been starved down to forty-five pounds and was within days, perhaps hours, of death from starvation when she was examined in Toronto in August 1922 and was brought back to life on insulin. Its effect on her body, she wrote her mother that autumn, was "unspeakably wonderful."[16] Fifty-eight years later, Elizabeth Evans Hughes Gossett, mother of three children, celebrated her fiftieth wedding anniversary. About that time I discovered that Elizabeth, who had been Banting's prize patient, was still alive, and I had the privilege of meeting her and learning of her life before and after the coming of insulin.

James Havens, who along with his family had given up and who wanted to die, was restored to life on insulin and by 1924 was up on skis; he lived for thirty-eight years on insulin. In 1923 Jim Havens wrote to Banting about having been allowed to eat egg on toast on Thanksgiving Day, and how it was his idea of the only food necessary in heaven. His physician wrote of Havens, "The restoration of this patient to his present state of health is an achievement difficult to record in temperate language. Few recoveries from impending death more dramatic than this have ever been witnessed by a physician."[17] Another American child, Ted Ryder, who was saved from certain death and fattened up on insulin, lived to become the first human to survive seventy years on insulin. He visited Toronto in 1990 to help celebrate the opening of a historical display about the discovery of insulin, and at his death in 1993 left the residue of his estate to support medical research at the University of Toronto.

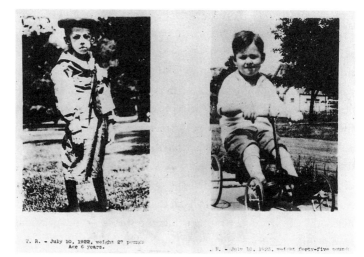

T. R. - July 10, 1922, weight 27 pounds
Age 6 years.

. R. - July 10, 1923, weight forty-five pounds

FIGURE 21A

Ted Ryder, before and after insulin, 1922-23

Fisher Library, University of Toronto

FIGURE 21B

Ted Ryder, Toronto, 1990. The power and glory modern medicine.

Author's collection

*

Many of the medical scientists of that era were still close to their Christian heritage, as Osler had been. Dazzled by the events they were witnessing with their diabetic patients—modern medicine's spectacular ability to affect the human condition—they drew on inherited concepts such as "miracle" and "resurrection" to describe the therapeutic impact of insulin. Patients talked about being reborn. The notion of salvation suddenly pervaded diabetes wards. Some patients even compared the physical sensations of their first insulin injections with being born again by the Holy Spirit at Pentecost. To see a diabetic patient brought out of coma by a single injection of insulin was to witness something akin to the resurrection of Lazarus. After they undeservedly won a brief and vicious battle for credit as the discoverers of insulin,*

---

*By the time of clinical trials on Leonard Thompson in January 1922, Banting had begun to believe that the senior scientists, Macleod and Collip, were appropriating his and Best's work. Reacting in kind to Banting's paranoia, Collip refused to tell Banting and Best how he made the critical purification of the extract, sparking an incident of violence in the lab reminiscent of how Canadian ice hockey players relieve their stress. Best had to pull Banting and Collip apart.

The superficial harmony of the research team was restored, but personal relations were deeply shattered. Banting came to believe that he had discovered insulin, with a bit of help from Best, working on dogs in the summer of 1921, and that Macleod and Collip had only helped somewhat in insulin's development. Macleod and Collip believed that they had made vital contributions to the research begun by Banting and that the discovery process included their work. In *The Discovery of Insulin*, I support the latter view. Collip's name probably should have been on the 1923 Nobel Prize. Charles Best's many later distorted accounts of the insulin research were a particularly egregious attempt to rewrite medical history and can be most charitably attributed to the effects of clinical depression. Another important chapter in the development of insulin involved the impressive cross-border collaboration of the University of Toronto with Eli Lilly and Company of Indianapolis, Indiana, to make insulin commercially available throughout North America with remarkable quickness and efficiency.

Frederick Banting and Charles Best were idolized for the rest of their lives as deities, the bringers of salvation, in the diabetes world—receiving wave after wave of adulation, more than was healthy.

One of Osler's kindred spirits and America's greatest diabetologists, Elliott Joslin, put the discovery of insulin in vivid scriptural context, citing the Old Testament as he talked about having witnessed so many near resurrections that he had seen enacted before his very eyes, Ezekiel's visit to the valley of dry bones:

. . . and behold, there were very many in the open valley; and, lo, they were very dry.

And he said unto me, Son of Man, can these bones live?

And . . . lo, the sinews and the flesh came upon them and the skin covered them above: but there was no breath in them.

Then said He unto me, "Prophesy unto the wind, prophesy, Son of Man, and say to the wind, Thus saith the Lord God: 'Come from the four winds, O breath, and breathe upon these slain, that they may live.'"

So I prophesied as he commanded me, and the breath came into them, and they lived, and stood up upon their feet, an exceeding great army.[18]

*

It had happened that the baby William Banting was baptized by the Reverend Featherstone Osler in Bond Head, Ontario, on the same day in 1849 that he baptized his own son, William. William Osler died in 1919, of empyema following a case of the Spanish flu, so he never lived to see the treatment of diabetes put on a sure foundation as a result of an idea that occurred to William Banting's son, Fred, working in the great educational and research complex in "Otnorot" that Osler's own career had inspired. He never lived to see children resurrected from the death by diabetes. Had he lived, Osler would not have been surprised that the promise of medicine was again being fulfilled, and of course he would have been delighted at the Canadian, the Toronto, and the Bond Head connection. Osler would have commented on how magically medicated that baptismal fluid must have been in 1849, and he might well have cited a line from his favorite author, Sir Thomas Browne, who writes in his classic work, *Religio Medici*, "Thus I call the effects of Nature the works of God."[19]

Alfred Nobel's instruction in his will that one of his prizes should celebrate the most important discovery of the previous year in physiology or medicine has proven almost impossible to honor. It often takes many years to evaluate medical breakthroughs, for they are usually complicated, cumulative, and have a delayed clinical impact. In the case of insulin, there was no doubt. The Toronto researchers were honored astonishingly quickly with the 1923 prize. It was awarded to Banting and Macleod. The two laureates, who intensely disliked each other, divided their prize money equally with Best and Collip. Llewellys Barker, the Canadian who had succeeded Osler as Professor of Medicine at Johns Hopkins, commented sagely at Toronto's Nobel Prize dinner, "In insulin there is glory enough for all."[20]

EPILOGUE: THE COLLAPSE OF LIFE EXPECTANCY

After the insulin breakthrough, which stood in especially brilliant contrast to the preceding years of war and slaughter, the potential of modern scientific medicine was no longer seriously questioned in the affluent Western countries. The North American 1920s are misunderstood in popular history as a frothy, frivolous decade of booze and broads, Babe Ruth and Al Capone. They were actually a time when doctors and their temples of healing had become new cultural heroes. Frederick Banting, for example, was the most famous person in Canada, popularly expected to conquer a new disease every month or two.* Doctors were the only people that the deeply cynical American social critic

---

*A British physiologist was once asked by a reporter from one of London's tabloid newspapers if he could confirm the story from Canada that the great Dr. Banting had finally discovered the cure for metabolism.

H. L. Mencken unqualifiedly hero-worshipped. The same can be said of Ernest Hemingway, a doctor's son, and of F. Scott Fitzgerald. In 1926, a year after the circus of the Scopes trial about Christian fundamentalism and evolution, the American Pulitzer Prize in fiction was awarded to Sinclair Lewis, a harsh critic of American business values and religious credulousness, for his novel *Arrowsmith*, glorifying medical humanitarianism and research. The Pulitzer Prize in nonfiction that year went to Harvey Cushing for his two-volume *Life of William Osler*. In 1931 Edith Gittings Reid's adulatory biography of Osler, *The Great Physician*, went through five printings in five months.

In Alberta, Canada, by the mid-1920s, Professor J. B. Collip had become one of the leaders in the isolation of parathyroid hormones. By the end of the decade, he had moved to McGill, where he created one of North America's first great modern research labs. In the United States in the mid-1920s, a team of researchers that included an insulin-dependent diabetic, George Minot, developed a therapy for pernicious anemia that garnered another Nobel Prize. In Boston Harvey Cushing and his students, inspired in part by the discovery of insulin — as were many other researchers in the 1920s — revived their interest in pituitary hormones, only to be beaten in the quest for human growth hormone by Herbert Evans in California. One of Cushing's protégés, Elliott Cutler, was beginning to explore what Cushing labeled as the new Northwest passage of surgery, the repair of the heart.

In towns and cities throughout North America, the 1920s saw the building and rebuilding of hospitals as centers of health care for whole communities, not just the indigent. Private patients were willing to pay for their own rooms and for access to skilled specialists and advanced medical technology. Mothers now gave birth in hospitals more often than at home. The best medical schools produced more and better graduates than ever before; many of the lesser schools, especially those catering to the old sectarian doctors, just closed their doors. By the end of the 1920s, the elite schools were turning away large numbers of applicants, a new phenomenon. Medicine was a good, a respected, and now a fairly lucrative profession to enter. By contrast, the mainstream Christian clergy were floundering in a swamp of postwar disillusionment, changing morals, disbelief, empty pews, and declining social status.

The 1930s were a decade of continued and spectacular medical advance even in the teeth of global economic hardship and in a climate of mounting fear, ignorance, and fatalism about other human events. The Canadian symbol of medical progress in that era was Wilder Penfield's well-publicized neurosurgical work, carrying on the Cushing tradition, at the Montreal Neurological Institute, affiliated with McGill University. Even as their society was degenerating into barbarism, German chemists in a reprise of their country's nineteenth-century research prowess developed the first generation of pharmaceutical compounds, the sulfonamides, that actually killed bacteria and cured disease. From Britain and America, penicillin and a host of antibiotics soon followed. A great struggle for mastery had effectively been won.

Then in the twenty years after the end of World War II and in the decades afterward, the progress of Western medicine was so great that the principal issue changed. Medicine's benefits to patients were so obvious and so important that the central problem became how to afford to pay for all that was emerging from the cornucopia.

Even if we pass over the ways in which medicine and medical research were perverted during the Second World War, especially in Nazi death camps, we cannot forget the ironies abounding in this story, as they lurk in most human events. As miraculous and wonderful and history-changing as the discovery of insulin may have been, its administration was not and is not a cure for diabetes. It is a maintenance therapy, and an imperfect one at that. Since the discovery of insulin, the total amount of diabetes in the human population has increased hugely, largely because of a tsunami-like surge of late onset, or type 2, diabetes caused in part by the increased affluence of our times. In its various forms, diabetes mellitus is a greater menace to human health now than it was in the days of Osler, Banting, Best, Collip, and Macleod. Strictly speaking, there is still no cure.

Still, diabetes is treatable. It is not clear whether there is yet any effective treatment for the destruction caused when the smallpox virus invades a human subject, any treatment beyond comforting a patient, much as Osler comforted his patient by reading the Bible and saying prayers in 1874. But that hardly matters, because in the years after 1885 when Montreal was ravaged by smallpox and fear, ignorance and fatalism, public health leadership and enlightened public opinion gradually won the war to immunize the vulnerable through vaccination while stamping out smallpox fires through isolation and quarantine. In the 1970s the World Health Organization administered the first successful campaign in human history to eradicate a disease. Driven from its last hideout in the deserts of Somalia and Ethiopia, the smallpox virus died.[1] The world has not seen a case of smallpox since 1978, and few of our children now bear even the stigmata of vaccination scars. Poliomyelitis, another virus controllable by vaccination, will be the next to go.

One of William Osler's greatest mistakes as a physician was to believe that humans live for a physiologically fixed period, about the biblical seventy years that he himself would happen to survive. Osler did not foresee the great payoff of modern medicine, which has been the expansion of human life expectancy by approximately thirty years since his heyday, an extension of our life span also enriched by better health at every stage, including old age. Osler did understand that health care—the work of physicians and a myriad of caregivers, scientists, public health workers, and others—is one of humanity's greatest success stories.

Question: If we are so successful at bringing good health to populations, why does our health care keep costing us more and more, almost as though we were getting sicker and sicker? After all, as we prevent fires from ravaging our community, we save on fire insurance. Where are the savings from preventing, say, the ravages of smallpox, quick death from diabetes, paralysis from polio, wastage from lung cancers and heart disease?

The answer is twofold. Generally, the benefits of the good health and longevity we enjoy are an enormous asset to humanity, the monetary value of which we cannot quantify and compare with the costs of our health care system. Secondly, the sad limit of the human condition is that health improvements are something of a mug's game. We master one disease—in fact, garnering great savings, say, in smallpox control—only to set us up eventually to suffer from another disease. Assuming we avoid accidental death, suicide, or being murdered, it is still the case that all of us, every single one of us, will eventually be defeated and killed by ill health. There is no mastery, no conquest of death. In the long run, the body always lets us down. There is no victory over the human mortality rate, which is stuck at 100 percent, give or take one's views about people as diverse as Jesus and Elvis. In this situation, the best analogy to explain the costs of health care, I believe, is to think of the struggles and expense we might go to in order to try and stop a snowman from melting, first during winter thaws, then in the growing warmth of spring. The more success we have, the tougher the struggle, the higher the costs.

By the twenty-first century, our successes with modern medicine have become so great that a modern Osler would be correct in assuming that we are bumping up against physiologically fixed limits to human longevity. The fixed period may not be sixty or seventy years, but it cannot be much more than one hundred. As we approach that limit, a number of health policy-makers, physicians, and others of goodwill are suggesting that we should consider the advice Osler gave facetiously a century ago in his famous address "The Fixed Period" and all agree to go gently (and relatively cheaply) into that good night by accepting one form or another of de facto euthanasia.[2]

That is not likely to happen in secular societies. There is a remarkable catch-22, perhaps the ultimate irony, involved in our substitution of faith in the healing powers of health care for the fatalism of traditional religious acceptance. When they faced sickness and death during the smallpox epidemic of 1885, many of the poor people of Montreal—like many of our own grandparents and perhaps parents, and like William Osler's clergyman father—were able to find consolation in the thought that the souls God was taking were going to another and a better life. Their life span would be eternal. Compare that with the best life span the new gods of health and healing hold out to us—a mere eighty or ninety or one hundred years. In secular societies the anticipated human life span shrinks catastrophically, a fact that does not please us and that we find almost impossible to accept with equanimity. Whether or not we rage, rage against the dying of the light, most of us are not inclined to go gently. We want to do everything we can to buy more time.

There are many other ironies and contradictions involved in the modern secular faith in health care, not least our discomfort in seeing flawed physicians as somehow godlike. They are not gods, they cannot be gods, and we know it, and we resent them for their imperfections; and our frustration is all the greater in our realization that they have no ultimate salvation to offer, no immortality. We have faith and they always let us down, not because they are getting worse at what they do—by any measurement they are constantly getting better and better—but mainly because we keep on raising expectations. They will never be able to meet our ultimate expectation, to be forever young and healthy.

Even as we understand this and rail against the problems of our modern health care systems, we still hope desperately for more breakthroughs, anything that will buy us a little more time before our personal world comes to an end at the moment of our death. Thus we lavish support on biomedical research, research being the closest we have to sacred causes in secular society. In the twenty-first century, we are even beginning to try and master the aging process itself, the ultimate goal being something like personal immortality. Woody Allen once said that he wasn't craving immortality through his movies; instead he wanted to achieve immortality by living forever. Perhaps no one should even consider having that much hubris, but there are still many millions of people in the world, including insulin-dependent diabetics, who are looking to health care to at least get them up to something approaching the normal life span their bodies are trying to deny to them. Around the world, human life is becoming less cheap, more valued. In his great lay sermon "Man's Redemption of Man," Osler urged health care providers to work for "the day when a man's life shall be more precious than gold."[3]

*

Human history is often a tale of woe and misery and evil, but not always. Medical history is often a tale of setbacks, quackery, misery, and, ultimately, always death. There is no shortage of critics of every aspect of our current health care systems, critics who point to problems ranging from our shifting definitions of illness through our therapeutic shortcomings, the growth of resistance to antibiotics, the menace of deadly modern bugs like HIV and the SARS virus, the (extremely implausible) possibility of smallpox being resurrected as a terrorist weapon, and the enormously complex problem of how to pay for all we can now do to prolong and improve human lives.

But that is what health care now does. Instead of just diagnosing and predicting the situation of patients, it now prolongs and improves our lives. Instead of forecasting the human climate, health care sometimes changes the human climate. We would not highlight the problems of today were we not able to take virtually for granted the progress of the past. You do not write manuals on how great machines break down and fail until after you have discovered how to make great machines. If asked when in history they would most prefer to be alive, most people would say the present, because it is an age when medical mastery is an everyday occurrence and fears and fatalism have been redefined almost beyond recognition.

In the last few years of the nineteenth century and the early decades of the twentieth century, the people of Canada and the United States were witnesses to some of the terrifying and inspiring events in the development of the health care system's ability to alter and improve our human condition. Within the span of a very few years, they came to have faith in the power of health care to save and extend our lives, a great and welcome new revelation. The new faith and power was shared by the people of Great Britain and Europe, and nowadays extends through most of the world. Thus we call the effects of nature the works of God.

Notes

These essays are based on the extensive primary and secondary research that went into the writing and arguments contained in four of the author's books:

- *Plague: A Story of Smallpox in Montreal*. Toronto: Harper-Collins, 1991; reissued with a new preface as *Plague: How Smallpox Devastated Montreal*. Toronto: HarperPerennial, 2003.
- *William Osler: A Life in Medicine*. Toronto: University of Toronto Press; New York: Oxford University Press, 1999.
- *Harvey Cushing: A Life in Surgery*. Toronto: University of Toronto Press; New York: Oxford University Press, 2005.
- *The Discovery of Insulin*. Toronto: McClelland & Stewart, 1982; Chicago: University of Chicago Press, 1983; rev. 25th

anniversary ed., Toronto: University of Toronto Press and Chicago: University of Chicago Press, 2007.

I also draw occasionally on a fifth book:

· *Banting: A Biography*. Toronto: McClelland & Stewart, 1984; 2nd ed. with new preface. Toronto: University of Toronto Press, 1992.

The notes contain the sources for all direct quotations used in the text, as well as references to certain other key secondary works. Most notes also refer to relevant portions of *Plague, Osler, Cushing, and Discovery of Insulin.* Much more elaborate bibliographies and citations are contained in these books. Readers with queries about my research or comments about this text should contact me by e-mail, m.bliss@sympatico.ca.

INTRODUCTION

1. Michael Bliss, "Growth, Progress, and the Quest for Salvation: Confessions of a Medical Historian," *Ars Medica: A Journal of Medicine, the Arts, and Humanities* 1, no. 1 (Fall 2004): 4–14; reprinted in *Essays in Honour of Michael Bliss: Figuring the Social,* ed. E. A. Heaman, Alison Li, and Shelley McKellar (Toronto: University of Toronto Press, 2008), 41–49.

CHAPTER ONE

1. William Osler, "Haemorrhagic Small-Pox," *Canadian Medical and Surgical Journal* 5 (1877): 289–304; *Osler,* 83.

2. *Plague,* 29–39. The best histories of smallpox are Donald R. Hopkins, *Princes and Peasants: Smallpox in History* (Chicago: University of Chicago Press, 1983); and F. Fenner, D. A. Henderson, I. Arita, Z. Jezek, and I. D. Ladnyi, *Smallpox and Its Eradication* (Geneva: World

Health Organization, 1988). Recent scholarship on the spread of vaccination can be found in the special issue "Reassessing Smallpox Vaccination," *Bulletin of the History of Medicine* 83, no. 1 (Spring 2009).

3. All details of the epidemic are drawn from *Plague*.

4. William Osler, *The Principles and Practice of Medicine* (New York: D. Appleton, 1892), 47.

5. *Montreal Star*, February 25, 1886; *Plague*, 262.

6. *Montreal Star*, February 1, 1886; *Montreal Gazette*, February 1, 1886; *Plague*, 257.

CHAPTER TWO

1. *Records of the Lives of Ellen Free Pickton and Featherstone Lake Osler* (Oxford: Oxford University Press for private circulation, 1915), 41; *Osler*, 22.

2. Harvey Cushing, *The Life of William Osler*, 2 vols. (Oxford: Clarendon Press, 1925), I:215–16; *Osler*, 121.

3. For a sweeping statement of this argument, see Richard Harrison Shryock's classic history, *The Development of Modern Medicine: An Interpretation of the Social and Scientific Factors Involved* (1936; reprint, New York: Knopf, 1947).

4. See Kenneth Ludmerer, *Learning to Heal: The Development of American Medical Education* (New York: Basic Books, 1985).

5. H. L. Mencken, *Happy Days, 1880–1892* (1936; reprint, Baltimore: Johns Hopkins University Press, 1996), 153–54, 284–96; *Osler*, 176.

6. Harvey Cushing, "Laparotomy for Intestinal Performation in Typhoid Fever: A Report of Four Cases . . ." *Johns Hopkins Hospital Bulletin* 9 (November 1898): 257–69; *Cushing*, 110.

7. Undated *New York Journal American* clipping, Cushing Papers, Yale University, reel 127, p. 582; also Cushing notes on Newcomb in John Fulton, *Harvey Cushing: A Biography* (Springfield, IL: Charles C. Thomas, 1946), 263–65.

8. Harvey Cushing, "A Method of Total Extirpation of the Gasserian Ganglion for Trigeminal Neuralgia . . ." *Journal of the American Medical Association* (1900): 1035; *Cushing*, 126.

9. Walker Evans to Harvey Cushing, December 18, 1907, Harvey Cushing Papers, Yale University Library, microfilm reel 127, p. 1541; *Cushing*, 126.

10. The extent of Halsted's addiction was only revealed in 1969 when Osler's manuscript "The Inner History of the Johns Hopkins Hospital" was published in the *Johns Hopkins Medical Journal* 125 (1969): 184–94; *Osler*, 211–12; *Cushing*, 116. For Halsted's career, see Gerald Imber, *Genius on the Edge: The Bizarre Double Life of Dr. William Stewart Halsted* (New York: Kaplan, 2010).

11. Harvey Cushing to Grace Osler, August 22, 1922, Cushing Papers, Yale University, reel 83, folder 577; *Osler*, 370.

12. Elizabeth Thies to Harvey Cushing, May 10, 1920, Cushing Papers, Yale University, folder 602; *Osler*, 337.

13. William Osler, *"Whole-Time Clinical Professors," A Letter to President Remsen Johns Hopkins University*, pamphlet (Oxford, 1911); *Osler*, 388.

14. Harvey Cushing, "Consecratio Medici," in *Consecratio Medici and Other Papers* (Boston: Little Brown, 1940); *Cushing*, 402.

15. William Osler, "Medicine in the Nineteenth Century" (1901) in *Aequanimitas*, 2nd ed. (Philadelphia: Blakiston's, 1922), 230; Osler, *Man's Redemption of Man*, address at the University of Edinburgh, July 1910; pamphlet (London: Constable, 1910); *Osler*, 393–94.

16. William Osler, "Unity, Peace, and Concord," in *Aequanimitas*, 447–65; *Osler*, 321.

17. Osler, *Man's Redemption of Man*.

18. William Osler, "The Faith that Heals," *British Medical Journal*, June 18, 1910, 1470–72; *Osler*, 276.

19. William Osler, *The Treatment of Disease: The Address in Medicine Before the Ontario Medical Association, Toronto, June 3, 1909*, pamphlet (London: Henry Frowde, Oxford University Press, 1909), 21.

20. E. G. Reid, "A Giver of Life," in Reid to Harvey Cushing, June 20, 1920, Cushing Papers, Yale University, folder 590.

21. William Osler, "Sporadic Cretinism in America," *American Jour-*

nal of the Medical Sciences 114 (October 1897): 378–401; *Osler*, 243–44. Osler seems to have given Hippocrates a totally imaginary daughter and invented her brave kiss.

22. Osler, *The Treatment of Disease*, 5.

23. Ibid, 8; *Osler*, 361.

CHAPTER THREE

1. "The Model Hospital," *Canadian Journal of Medical Science* 6 (1881): 154–56; *Osler*, 147. Osler's authorship of this anonymous article has occasionally been questioned.

2. Sandra F. McRae, "The 'Scientific Spirit' in Medicine at the University of Toronto, 1880–1910" (Ph.D. diss., University of Toronto, 1987), chap. 5; *Osler*, 251.

3. W. W. Francis to B. C. MacLean, November 13, 1951, W. W. Francis Papers, Osler Library, McGill University; *Osler*, 251.

4. Elizabeth Osler to Jennette Osler, December 8, 1880, Osler Papers, Fisher Library, University of Toronto; *Osler*, 31.

5. William Osler, "The Master-Word in Medicine," in *Aequanimitas*, 363; *Osler*, 299–300.

6. See Michael Bliss, *A Canadian Millionaire: The Life and Business Times of Sir Joseph Flavelle, 1859–1939* (Toronto: Macmillan, 1978).

7. In addition to *Discovery of Insulin*, see Michael Bliss, "J. J. R. Macleod and the Discovery of Insulin," *Quarterly Journal of Experimental Physiology* 74 (1989): 87–96; and Alison Li, *J. B. Collip and the Development of Medical Research in Canada* (Montreal: McGill-Queen's University Press, 2003).

8. F. G. Banting, Primary Insulin Notebook, Insulin Collections, Fisher Library, University of Toronto; *Discovery of Insulin*, 50.

9. *Discovery of Insulin*, 20.

10. James Havens Sr. to E. C. Gale, March 1, 1921, Havens Family Papers, privately held (copy in Bliss Papers, Insulin collections, Fisher Library, University of Toronto); *Discovery of Insulin*, 136.

11. John R. Williams, "A Clinical Study of the Effects of Insulin in

Severe Diabetes," *Journal of Metabolic Research* 2 (November 1922): 729; *Discovery of Insulin*, 136.

12. Israel Kleiner, "The Action of Intravenous Injections of Pancreas Emulsions in Experimental Diabetes," *Journal of Biological Chemistry* 40 (1919): 153-70; *Discovery of Insulin*, 40-42.

13. *Cushing*, 382-84.

14. William Osler, with the assistance of Thoms McCrae, *The Principles and Practice of Medicine*, 8th ed. (New York: D. Appleton, 1918), 435.

15. Harvey Cushing to Paul D. White, January 18, 1922, Cushing Papers, Yale University, microfilm reel 20, p. 1058; *Cushing*, 384.

16. Elizabeth Hughes to Antoinette Hughes, October 6, 1922, Elizabeth Hughes correspondence, Fisher Library, University of Toronto; *Discovery of Insulin*, 153.

17. Williams, "A Clinical Study," 734; *Discovery of Insulin*, 162.

18. Elliott Joslin address at the opening of the Lilly Research Laboratories, 1934, Eli Lilly and Company Archives, Indianapolis; Ezekiel 37:2-10; *Discovery of Insulin*, 164.

19. Sir Thomas Browne, *The Religio Medici and Other Writings*, Everyman ed. (London: Dent, 1906), 18.

20. *Toronto Daily Star*, November 27, 1923; *Discovery of Insulin*, 229-33.

EPILOGUE

1. For the campaign that killed smallpox, see Fenner et al., *Smallpox and Its Eradication*.

2. William Osler, "The Fixed Period," address at Johns Hopkins University, February 22, 1905, in *Aequanimitas*, 389-411. A correspondent wrote in the *Globe and Mail* on November 15, 2008: "Maybe the aboriginal legends were right—when you strike camp, you leave the very elderly who are sick and infirm in a nice spot by the river, preferably with a bottle of pinot noir."

3. Osler, *Man's Redemption of Man*, 34.

Index

acromegaly, 61
Addison's disease, 61
Aikins, William, 66
Allen, Frederick, 76-77
Allen, Woody, 93
American Osler Society, 41
antibiotics, 89
Arrowsmith (Lewis), 88

Banting, Frederick Grant, 72-85, 87
Banting, William, 84
Barker, Llewellys, 85
Beaugrand, Honoré, 18
bedside teaching, 43, 57

Bessey, William, 24
Best, Charles, 78-84
Brödel, Max, 49, 51-52
Browne, Sir Thomas, 85, 94

Campbell, Walter, 71
Canada, strength in medicine, 3-4, 67
Carrel, Alexis, 47, 51
Charlottetown, Prince Edward Island, 21
Christian Science, 54
clergy: decline of, 55, 88; and smallpox, 11, 27-30
Cleveland Clinic, 48

clinical clerkship, 43
Coderre, Joseph Emery, 23, 25
Collip, James B., 70–71, 78–80,
    83–84n, 88
cretinism, 58–59, 74
Crile, George, 47–48
Cushing, Harvey, 43–51, 61,
    77–78, 88
Cutler, Elliott, 88

diabetes, 61, 72–85, 89–90
Discovery of Insulin, The (Bliss),
    79, 83n

Eaton, Timothy, 70, 71
Eli Lilly and Company, 83n
Endocrine Society, 77
endocrinology, 47, 74, 77–78
erysipelas, 14, 23
euthanasia, 92
Evans, Herbert, 88
Ezekiel, 84

faith, 3; and disease, 9–10,
    25–30, 92; in physicians,
    55, 92
faith healing, 34, 45, 54–55
fatalism, 9–10, 28–29, 41
Fitzgerald, F. Scott, 88
Flavelle, Joseph Wesley, 68, 70
Flexner Report, 41, 70

Gates, Rev. Frederick, 50–51
German influences, 43, 89
Goodman, Joanne, 3, 4

Graham, Duncan, 71
Great Physician, The (Reid), 88

Halsted, W. S., 41, 42, 50
Harvard University, 37, 44, 47, 76
Havens, James, 76, 81
Hemingway, Ernest, 88
Hippocrates, daughter of, 60, 99
HIV, 93
homeopaths, 25
Hopkins, Johns, 39. See also
    Johns Hopkins institutions
Hospital for Sick Children,
    Toronto, 72
Hôtel-Dieu Hospital, Mon-
    treal, 13
Hughes, Elizabeth, 81
human growth hormone, 88
hydropathy, 25, 37

insulin: controversy, 83–84n; dis-
    covery of, 75–85; limitations
    of, 89–90

Jenner, Edward, 11, 13, 23, 36
Johns Hopkins institutions,
    37–50; and Canada, 66–71
Joslin, Elliott, 84

Kelly, Howard, 41, 42
Kleiner, Israel, 77
Koch, Robert, 58

Labatt's beer, 25–26,
Lewis, Sinclair, 88

life expectancy, 90–92
Life of William Osler (Cushing), 88
Longley, George, 13

Macallum, A. B., 66, 67
Macleod, James J. R., 71–72,
    79–80, 83–84n
Mallam, Patrick, 58n
Massachusetts General Hospital,
    44
Mayo Clinic, 47
McGill University, 34–35, 37, 43,
    65–67, 88–89
McRae, Sandra, 67
medical education, moderniza-
    tion of, 36–43, 88–89
Mencken, H. L., 41, 88
Minkowski, Oskar, 74, 78
Minot, George, 88
Montreal, 7–30, 65–66, 90
Montreal General Hospital,
    13, 35
Montreal Neurological Insti-
    tute, 89
myxedema, 58, 74

neurosurgery, 45–51
Newcomb, Simon, 44
Nobel Prizes, 47, 83–84n, 88
North Atlantic Triangle, 4, 60
nursing, 36

Ontario Medical Association, 61
Osler, Rev. Featherstone, 33–35,
    84, 92

Osler, Sir William, 4, 6, 42, 52,
    57; baptism and death, 84–85;
    career and influence, 33–61; on
    curing, 54–61; on diabetes, 61,
    74–75, 78; and fixed period,
    90–92; influence in Canada,
    65–70; "Man's Redemption of
    Man," 93; on medical prog-
    ress, 53–54, 61; and smallpox,
    7–8, 11, 21, 22n, 29, 35
Osler boys, 67
Oxford University, 51, 56, 70

Paget, Stephen, 53
parathyroid hormone, 88
Pasteur, Louis, 36
patent medicines, 25, 26
Penfield, Wilder, 89
penicillin, 89
pernicious anemia, 88
Peter Bent Brigham Hospi-
    tal, 47
philanthropy, 39, 51. *See also*
    Rockefeller philanthropy
pituitary, 47, 49, 88
Poe, Edgar Allan, 18
poliomyelitis, 90
*Principles and Practice of Medicine,
    The* (Osler), 21, 39, 43, 50, 78
Pulitzer Prizes, 88

Reid, Edith Gittings, 56, 88
Religio Medici (Browne), 85, 94
residency training, 43
Riel, Louis, 15, 24

Rockefeller philanthropy, 51, 70, 71, 72, 76-77

Ross, Alexander Milton, 23, 25, 36

Ryder, Ted, 81-82

sanitation, 10-11, 24

SARS, 93

smallpox, 7-30; eradication of, 90, 93

snowman metaphor, 91

sulfonamides, 89

surgery, 36, 44-51

Surgery, Gynecology and Obstetrics, 72

Thompson, Leonard, 78, 81, 83n

thyroid, 58-60

Toronto, rise of, 66

Toronto General Hospital, 68-70, 78

Toronto School of Medicine, 66

trigeminal neuralgia (tic douloureux), 45-46

ultramontanism, 28

United States, leadership of, 3, 47-48

University of Alberta, 71-72, 88

University of Edinburgh, 43

University of Pennsylvania, 37

University of Toronto, 66-85

University of Western Ontario, vii, 3, 72

vaccination, 11-30, 36, 90

Welch, William, 41, 42

Western Reserve University, 71

Wilson, Daniel, 66

World Health Organization, 90

Wright, Ramsay, 66, 67